The Meaning of Consciousness

Studies in Literature and Science
published in association with the
Society for Literature and Science

Titles in the series

Transgressive Readings: The Texts of Franz Kafka and Max Planck
by Valerie D. Greenberg

A Blessed Rage for Order: Deconstructionism, Evolution, and Chaos
by Alexander J. Argyros

Of Two Minds: Hypertext Pedagogy and Poetics
by Michael Joyce

The Artificial Paradise: Science Fiction and American Reality
Sharona Ben-Tov

Conversations on Science, Culture, and Time
by Michel Serres with Bruno Latour

Genesis by Michel Serres

The Natural Contract by Michel Serres

Dora Marsden and Early Modernism: Gender, Individualism, Science by Bruce Clarke

The Meaning of Consciousness by Andrew Lohrey

The Troubadour of Knowledge by Michel Serres

The Meaning of Consciousness

Andrew Lohrey

Foreword by N. Katherine Hayles

Ann Arbor

THE UNIVERSITY OF MICHIGAN PRESS

Published in the United States of America by
The University of Michigan Press
Manufactured in the United States of America
⊗ Printed on acid-free paper

2000 1999 1998 1997 4 3 2 1

A CIP catalog record for this book is available from the British Library

Library of Congress Cataloging-in-Publication Data

Lohrey, Andrew.
The meaning of consciousness / Andrew Lohrey.
p. cm. — (Studies in literature and science)
Includes bibliographical references and index.
Contents: Approaches and methods — Types of relations — Nonsymmetrical emergence — Substance and subjectivity — Meaning and consciousness — Freud and Matte-Blanco — Paradigms of consciousness — Awareness and creativity — Unity of mind and body — The isomorphic sign — Signifiers and signifieds — Discourse — Coda — Notes — Glossary — Bibliography — Index.
ISBN 0-472-10821-2 (alk. paper)
1. Consciousness. 2. Symmetry. I. Title. II. Series.
B808.9.L64 1997
126—dc21 97-4581
CIP

To Ammachi,
mother of love

Acknowledgments

This book has greatly benefited from various discussions, correspondence and support from a range of people. I would like to thank all those who have so generously given their time, experience and assistance to this enterprise.

I am indebted to the personal support I received from Rus Joyner who gave me the confidence to begin this book in 1980. I also thank Gunther Kress who provided me with the academic shelter of a doctoral program which assisted in bringing this work to a completion and in this regard I acknowledge the assistance of Stephen Muecke and Helen Irving, my academic supervisors. I am also grateful to Meaghan Morris who taught me semiotics many years ago.

I am deeply indebted to N. Katherine Hayles whose encouragement, support and criticism nourished and gave coherence to the isomorphic concepts used here.

The many drafts of this book have benefited from the comments, criticisms and advice I have received from several people. In particular I thank Venetia Somerset, Graeme Smith, Phillip Howard and my wife Amanda, whose has been a continual source of intelligence throughout this long period of research and writing. I am deeply grateful to Amanda for her unfailing inspiration, tolerance and the help she has given me in so many ways over this period.

Contents

Foreword

The Meaning of Consciousness is not a timid book. Styling itself nonmaterial and nonrationalist, it nevertheless aims to give an account of consciousness that can lay the foundation for a more scientific treatment of consciousness than has hitherto been the case. In a sense, it can be understood as returning to the schism that took place in the seventeenth century when the alchemists were discredited by the emerging power of the Royal Society. Whereas the alchemists thought the universe was animate and made no clear distinction between spirit and matter, the chemists who were their successors argued that the material world was governed by physical laws that should be understood as operating invariantly, independent of whatever spiritual value one might attribute to them. Andrew Lohrey wants to go back to this crossroads and take the other fork. He starts from the premise that the world is animate, and he searches for a way to place consciousness within this living universe as an embedded phenomenon.

At the center of his analysis is the idea of isomorphism. Emphasizing contextual embeddedness, he traces three contexts that operate isomorphically with one another. The largest is the "cosmic consciousness from which all material existence emerges." Nested inside this is the context of individual subjectivity, which Lohrey interprets as "an aspect of the cosmic context operating through the interactions of the nervous system." Then emerging from subjectivity is the context of discourse, operating as a "manifestation of cultural consciousness." These three contexts are codependent and coevolving. Rather than one affecting another through linear causality, they are bound together through recursive feedback loops. Interactions between them are governed by symmetry, along with its necessary correlates, asymmetry and nonsymmetry. The feedback loops have the effect of correlating symmetry operations within one context with symmetry operations in another, so that they are in constant dynamic interaction with one another.

We can understand this in a commonsense way by noticing how Lohrey negotiates the social constructivist/realist divide. He agrees with the constructivists that we never see the world from an impossible God's eye view independent of human perception and neural processing. Rather, he acknowledges that we construct the world in the very process of perceiving it. But unlike Humberto Maturana and Francisco Varela, who see the environment acting merely as a "trigger" for reactions determined solely by the internal organization of a system, Lohrey stresses the coevolution and coadaptation of the human nervous system with its environment. He points out, for example, that if our environment did not reflect light of varying wavelengths, we would not have evolved color vision. This potential within the environment stimulates the development of color vision, and color vision stimulates humans to modify their environment so that it has even more potential for colors than it did before. Using this holistic approach, Lohrey systematically develops a perspective radically different from a viewpoint based on a presumed separation between subject and object. Instead of seeing a world "out there" separate from an observer, Lohrey emphasizes the mutual coconstruction of world, subject, and discourse.

Often this perspective yields startling new insights. To illustrate, consider his critique of Saussean linguistics. He finds misguided its emphasis on differences among phonetic units rather than similarities between them. Such an emphasis, he suggests, is typical of approaches that concentrate on conscious awareness but remain oblivious to preconscious modes of perception and action. To illustrate how similarities are also important to meaning production, he points out that the physical production of sound relies on perceiving one "t," for example, as similar to other "t's." When the vocal apparatus produces the sound "t," it can do so only remembering the actions that produced similar sounds before and repeating them. Thus, from a bodily perspective, similarity provides the ground for the perception of differences. To concentrate only on difference at the expense of similarity is akin to listening for the sound of one hand clapping.

Written in the tradition of speculative philosophy, *The Meaning of Consciousness* draws upon a wide range of work in the sciences and humanities, including artificial life, chaos theory, David Bohm's theory of implicate order, deconstruction, linguistics, and critical theory. It aims to provide a new conceptual framework that does not reinscribe the subject-object dichotomy and that can relate the experience of human consciousness to isomorphic structures in culture and the natural world. Along the way, it provides a wealth of new insights, including pointing out recursive similarities between mind and environment and suggesting new ways to think about the problematics of defining system boundaries. Throughout, this text insists that the meaning of consciousness is nothing other than the embedded symmetries that connect it with the world. Even if readers are finally unpersuaded by the theories of consciousness put forward here, they will do much to instruct and delight them.

Dense, difficult, and closely reasoned, *The Meaning of Consciousness* is startlingly original. To find works comparable to it in scope, ambition, and ability to upset traditional apple carts, I think of such texts as Spencer-Brown's *The Laws of Form,* Maturana and Varela's *Autopoiesis and Cognition,* and Gregory *Bateson's Steps to an Ecology of Mind.* Finally, what is the meaning of *The Meaning of Consciousness?* Although no work this complex could possibly be summed up in a short phrase, a Shakespearian paraphrase probably comes as close to that impossible task as anything: context is all. The meaning of this book cannot be separated from the journey one takes through it. I urge you to take that journey.

N. Katherine Hayles

Introduction

Nearly one hundred years ago Sigmund Freud saw consciousness as a fact of individual experience lying open to immediate intuition. This fact, however, defied explanation or description. The inability to describe consciousness has persisted until the present with some philosophers; for example, Colin McGinn argues that "consciousness is beyond the reach of the human mind, that cognitively we are not equipped to understand it in a way we understand other phenomena."[1]

An explanation of consciousness appeared to be at hand when *Consciousness Explained,* by Daniel Dennett, was published in 1991. Yet Dennett confessed that his own explanation was to replace one set of Cartesian metaphors of the mind with metaphors of technology. For example, "the theatre," "the witness," and "the central meaner" were supplanted by "software," "virtual machine," and "multiple drafts."[2]

Another approach to the problem of consciousness has come from philosophers, such as John Searle, who, following Franz Brentano's lead, believe that consciousness is intentional; that is, any conscious state will be directed at something or other.[3] But if we accept this ambiguous concept of intentionality (a something is directed at a something else), unconscious states still remain unexplained. Searle has a difficulty with unconscious states and tries to overcome it by reducing unconsciousness to the physical processes of neurophysiology.[4] This reductionism amounts to an erasure of the unconscious, a crucial denial inherent, as I will argue, in a rational view of consciousness.

Over the last two decades cognitive science has provided us with several views and descriptions of mental processes that have been of little help in explaining consciousness. The basic approach of cognitive science is that cognition is a rule-based manipulation of symbols, that is, a symbolic computation of information. The most visible manifestation of cognitivism is in artificial intelligence.[5] Disputes in this field appear to center around the role and significance of artificial intelligence, questions of computation and noncomputation, and whether machines like the Turing machine can think. According to Hilary Putnam there is some pessimism (in writers like Dennett and Jerry Fodor) about the computational model and even about the general state of cognitive science.[6] In contrast to this relatively mechanical approach, Francisco Varela, Evan Thompson, and Eleanor Rosch, influenced by Eastern philosophy, have concentrated more on the processes of cognition and have given cognitive science a more organic definition. But even organic cognition can not provide us with a description of consciousness.

Yet interest in the nature of consciousness is growing. The September 1992 issue of *Scientific American* was given over entirely to articles exploring "Mind and Brain." Francis Crick and Christof Koch wrote a piece titled "The Problem of Consciousness." Its thesis was that consciousness can be explored by scientifically investigating the visual system. Both authors, however, declined to define consciousness, "because of the dangers of premature definition."[7]

All this activity indicates that while the problems of consciousness have historically defied Western descriptions, the challenge remains. Derek Denton, who presumably has not studied Eastern philosophy, has described the possibility of unraveling the nature of consciousness as the ultimate challenge to human ingenuity.[8]

Two features are common to the large amount of literature about consciousness that has appeared in recent years, as well as in the large conference "Towards a Science of Consciousness—Tucson II" in 1996. These are (1) a lack of any systematic theory, and (2) rational, materialist approaches that make mechanistic and instrumental attempts to unravel the mysteries of this essential field of "dark matter." These two factors are interlinked, for the failure to theorize consciousness is due to the limitations of a rationalist/materialist perspective, and all that that entails in terms of classical scientific method and experimentation.

In other words, a science of consciousness should rely, not on reason or empiricism and their attendant conventions, which ask how consciousness arises from matter, but on a perspective paradigm, which reverses this question to ask how matter arises from consciousness. If this seems a radical departure from a proper scientific approach, recall that quantum theory accepts nonmaterial energy fields as the underlying reality for the manifestation of particles. This metaphysical view represents the modern understanding of physical reality.

Yet what is known about quantum matter fields is not always carried through to questions about consciousness because, one presumes, the traditions of materialism are too strong. Another difficulty with a rationalist/materialist approach is that it automatically posits a rational consciousness as its objective. A rational consciousness is conceived as a self-evident, featureless state of explicit awareness, a state that can explain little about its own implicit, nonrational, and unconscious structures. If members of the scientific community concerned with consciousness are to unravel the problem, they must first accept that consciousness is not a rationally based phenomenon, in other words, that a possibility of description beyond self-evidence exists. Second, it will be necessary for us to set aside the traditional academic caution against "premature definitions" and risk a theory of consciousness. Such a theory would have to be universal and yet particular enough to account for any specific moment of consciousness, be it sensation, sign, or the perception of an environmental pattern or object.

The Meaning of Consciousness both develops and works from a new, nonrationalist and nonmaterialist framework or paradigm. The theory presented here challenges much within the scientific and Cartesian traditions of analysis and thought that have been around since 1619, when the Angel of Truth revealed in a vision to René Descartes that science should be based, like mathematics, on self-evident first principles. Some years later Isaac Newton deanimated the physical world with his *Mathematical Principles of Natural Philosophy*. The reinforcing effects of this deanimated, mechanical world based on self-evident first principles were a dualist and mechanical tradition that separated the physical from the metaphysical, spirituality from materialism, and that in some areas such as contemporary linguistics developed a formalism seen as separate from subjectivity. These three-hundred-year-old traditions, which have developed into the prescriptions of science, still serve most writers and researchers, yet they are manifestly inadequate when applied to subjectivity, meaning, discourse, and consciousness.

A nonrational, nonmaterialist framework that challenges these traditions leads to the present theory of consciousness, which is holistic, or more specifically *isomorphic,* that is, pertaining to the similarity of forms. This framework draws some of its strength from the philosophy and practice of close reading and discourse analysis. Discourse analysis is concerned with the rhetorical traditions and conventions of habit and the worldviews that emerge from these habits. The three-hundred-year-old traditions of science are supported by a set of "scientific" discourses that continually reinforce and perpetuate a reasoned materialism. To break with this limiting, so-called objective view is to break with the kinds of discourses that structure and perpetuate a deanimated, mechanical world. The present isomorphic theory of consciousness attempts to prosecute, with a high degree of fidelity, isomorphic discourses that serve and reinforce an animate, holistic worldview.

This is a radical break from "objective" scientific method that will most probably directly challenge those workers in the field of consciousness who rely upon the discursive traditions of a reasoned materialism. To indicate the depth of this challenge it is only necessary to state the general isomorphic propositions of this book. These are that consciousness has three general "layers": (1) a holistic context involving a cosmic consciousness from which all material existence emerges, (2) the context of individual subjectivity, an aspect of the cosmic context operating through the interactions of a nervous system, and (3) the context of discourse, which is the manifestation of cultural consciousness and which emerges from the context of subjectivity. These three contexts are integrated within each other and occupy a single continuous field of self-reflective meaning.

The integration of this single continuous field of meaning is constructed from the emergence of discourse from subjectivity, the emergence of subjectivity from cosmic consciousness, and the emergence of cosmic consciousness from the potentials of symmetry. The basic operation, or in the rhetoric of quantum science, the "wave function" that links the three "layers," is recursive symmetry. The foundational structural principle of this entire field is thus symmetry. Why is this a symmetry field, and why is consciousness based on symmetry? This is a question for which I find no easy answer; yet all the evidence from science, religion, and the humanities seems to suggest that this is the case.

A full understanding of consciousness entails a recognition of the integration of these three contexts. To ignore any one of them would result in an incoherent and decontextualized outcome, one that would only partially cover the subject. For some readers the notion of a cosmic consciousness may seem to import an element, not merely of the metaphysical, but of a religiosity not commonly associated with scientific approaches to mind and consciousness. It is part of my project to argue that the reason we have subjectivity at all is because it reflects the qualities of cosmic consciousness already existing in the universe.

Many theorists in the field of physics (Heisenberg, Eddington, Pauli, Schrödinger, De Broglie, Planck, and Bohm, to name a few) have been forced to confront the concept of an animate universe and have skirted around it in various ways. Flirting with the idea of God is possible if our focus is physics and our approach a materalist one. For example, David Bohm, one of the most distinguished theoretical physicists of this century, tells us, "one could say that through the human being, the universe is making a mirror to observe itself."[9]

Such flirtation with the notion of an animate universe, however, is not possible for a serious study of consciousness in which the approach is nonmateralist. We have to go beyond the merely suggested or implied, and the odd throwaway line to confront and begin to detail, in a systematic and coherent manner, the universal qualities of consciousness in all its cosmic and individual forms.

The three contexts of consciousness broadly form the subject matter for the three parts to this book. Part 1 deals with the holistic context of cosmic consciousness in relation to its semantic structure, which is based upon the universal, emergent relation of symmetry. Part 2 looks at the structure of subjectivity in terms of the same set of relations, while part 3 focuses on these same relations insofar as they structure discourse. The three contexts of consciousness therefore have the same recursive, semantic structure, and this structure runs through the entire field of meaning.

The present theory proposes that symmetry is a relationship, and as all relations are relations of meaning, and as meaning and consciousness are the same, consciousness represents a continuous, immanent, and formative field that encompasses and links everything within this undivided universe. Symmetry is therefore the cosmic, animate force behind the order and organization of the universe, an order that produces the emergence of the material world and the interactions of that materiality, an order that determines the environment as well as creating the structure and the content of subjectivity and discourse. The qualities of this universal, omnipresent, omniscient, animate force are contained within the symmetry potentials of relationships. This isomorphic theory is thus a nonrationalist and nonmaterialist theory of consciousness. It is nonmaterialist in that the structural components of the theory are relations, and relations are never material. It is nonrationalist in that the foundational relationships are those of symmetry, and symmetry eludes the reason of cause and effect.

Thus this nonrationalist and nonmaterialist isomorphic theory of consciousness links the physical to the metaphysical, science to spirituality, biology and psychology to semantics, and all knowledge to subjectivity. These integrations have some similarities with the pre-Cartesian, alchemical, unified view of the world that postulated a spiritual, cosmic order independent of our choice; an anima mundi or a world soul that sleeps in matter and is responsible for such things as the birth of new stars. This was a view held by Johannes Kepler, who at the beginning of the seventeenth century formulated his three famous laws of planetary motion, the pillars upon which Newton based his theory of gravitation. The similarity of an isomorphic theory to this pre-Cartesian view should not be pushed too far, however, for it exists mainly in the content of an undivided and animate universe. But even this proposal seriously challenges contemporary scientific predispositions, which have lived for so long with the deanimation of a Newtonian, mechanical world.

The isomorphic theory presented here also has some tentative connections with the view of cosmic intelligence held by the "process" theology propounded by John Cobb as well as with the nondual (advaita) traditions of Hinduism. Scholars of Indian philosophy may see some parallels with the ancient scriptures of the Vedas, particularly the Upanishads. I became aware of these connections only when this book was almost completed. While these links are interesting and generally supportive, this is not primarily a book about theology, or mysticism.

I draw attention to the connection with these nonmaterialistic world views, because such views are usually associated with superstition and nonscientific endeavor. *The Meaning of Consciousness* is not a work of superstition, nor is it nonscientific. I would call it a scientific theory of consciousness. It is scientific in its intellectual and systematic approach to a puzzle that has confounded many. It is, however, a nonrationalist and nonmaterialist theory, and if reason and materialism are identified with science, then it cannot be scientific. But such a pattern of identification is too narrow, too restrictive, and is in any case irrelevant to recent developments in "new physics."

In many ways *The Meaning of Consciousness* is a very Western and intellectual study of the meaning we find in the metaphor of the "heart," that is, the heart as empathy, compassion, and love. This is the meaning that binds, connects, and interrelates an undivided universe. Yet this book does not follow the path of Eastern mysticism, nor does it fit nicely into the Cartesian and Newtonian traditions of Western science and technology that are so evident in the present state of cognitive science.

To reiterate, the central principle in this isomorphic theory of consciousness is symmetry. Symmetry is a principle common to mathematics, complex systems, and science generally. Much of science has been devoted to discovering fundamental symmetries in the universe, such as Einstein's formula for the symmetry of mass and energy. The general principle of symmetry is, however, almost never applied in the social sciences, apparently because it does not fit well with formalist logic and the traditions of reason.

The benefits of an isomorphic perspective are already evident in "new physics" and the "science of complexity," which employ variations of isomorphism to calculate the chaotic behavior of complex systems. Yet the wonders of quantum mechanics and chaos theory have not, by themselves, produced a wider nonrationalist theory of consciousness, nor has the more general science of complexity, despite developments in these areas that offer many signposts to an isomorphic theory of consciousness. In this book I have attempted to use some of the more basic signposts and apply them to the field of meaning and consciousness. One of these is "recursive symmetry," which is the central organizing principle in all complex systems. Another is the notion that complexity arises from simplicity. Chris Langton tells us, "The science of Complexity teaches us that the complexity we see in the world is a result of underlying simplicity."[10] If this is the case, then we could expect that the complexity of consciousness will have an underlying simplicity. This is the approach taken throughout this book. The transformation of simplicity into complexity is the underlying theme of this isomorphic theory, for emerging from a simple, symmetrical structure is the whole complexity we know as the universe.

The present isomorphic theory proposes that individual subjectivity (the second "layer" of consciousness) is the central organizing system of all knowledge, meaning, and interpretation. This is a humanist perspective, but not a brand of nineteenth-century progressive humanism with its dream of "human engineering" and of control over nature. Rather, the humanism here is nonrationalist, an isomorphic humanism that privileges the implicit, unconscious, enfolded meaning that interconnects all things. Far from controlling nature, an isomorphic humanism views individual subjectivity as an embedded expression of the fundamental, formative, symmetrical processes already at work in the whole of the universe. It embeds us within nature rather than separating us from it.

A second distinctive feature of this approach is the expectation that subjectivity can be scientifically known. Traditional attempts to know subjectivity have come from two quarters; the "objective" and the introspective. Both approaches have failed to live up to expectations. Objectivism in science has always suffered from the difficulties that come from naive realism; for example, from the belief that the laws of nature are material laws that exist independently from discourse, culture subjectivity, and consciousness. Academic psychology in particular has suffered from this stance in its overwhelming adherence of the pragmatics of behaviorism. Behavioristic psychology infers mental life from behavioral evidence, which ultimately consists of the motion of material objects.

The introspectionists have not fared much better. The introspective world of subjectivity is said to be based on "qualia." Qualia are the collection of personal and subjective experiences, feelings, and sensations that produce "how things seem to us." The problem with the term qualia is that it represents isolated, unsystematic, unreliable information, a poor basis for a workable, scientific framework. More basically, *qualia* and the category of introspection always represent one part of a rational Cartesian pair. When we speak about qualia, we are always already implying the other half of the pair; the "objective," physical "out there." This kind of binarism is mechanical and reductive and inappropriate for the complexities of subjectivity and consciousness.

The present isomorphic approach assumes that the subjectivity of the scientific observer can be known scientifically, not "objectively" or "introspectively" but through a systematic model of the meaning of meaning. The meaning of meaning refers to the principle of symmetry that structures the continuous field of the three "layers" of consciousness. This isomorphic approach therefore connects the observer to the observed within a single field of meaning and consciousness.[11]

The present isomorphic view defines consciousness as an emergent, asynordinate, isomorphic phenomenon. To what do these terms refer? *Emergent* refers to the changes that occur from symmetry transformations, when, for example, a particle emerges from its underlying energy field, or more generally, when a context is transformed into a series of constituent features or, in reverse, when a constituent feature transforms into a context. These emergent transformations occur all the time at every level of the universe and in every context. The three large transformations that are relevant to our discussions on consciousness are the emergence of physical existence from cosmic consciousness, the emergence of conscious awareness from prereflective unconsciousness, and the emergence of discourse from subjectivity. In this sense, discourse represents the emergent, cultural forms in which consciousness is made manifest.

The overall structure of consciousness, however, is never and can never be fully explicit. Rather, it is *asynordinate.* The term was used by the physicist David Bohm, who coined it to refer to structures that have varying degrees of hidden implicitness. The image of an iceberg with the majority of its structure hidden beneath the water is asynordinate. The importance of the term reflects the general isomorphic perspective of the present approach. This approach proposes that unconscious meaning has a hidden, enfolded, implicit, and symmetrical structure, while conscious meaning has a visible, unfolded, explicit, and nonsymmetrical or differential character. Both of these aspects, the implicit and the explicit, the enfolded and unfolded, the hidden and the visible, are integrated so that the overall structure of consciousness is a combination of both: it is an asynordinate system.

The same can be said of discourse, for this too is an asynordinate phenomenon, being an integration of both implicit and explicit meanings. The same can be said of quantum matter fields. These are asynordinate because such fields are integrated systems of an implicit, symmetrical, and hidden field and explicit, differential particles. The particles and their field are never separate. In like manner, the explicit and differential meanings of discourse are never separate from their hidden foundational field of implicit meanings. Discourse and quantum behavior, traditionally seen as poles apart, are not so disparate, for these activities share the basic asynordinate structure that runs through the entire universe. This is the structure of consciousness.

Finally, consciousness is *isomorphic,* which means that the relations that provide its structure are circular, repetitious, reflective, holistic, and formative. Thus the overall architecture of the totality of the unified field of consciousness, which includes the three "layers" or contexts of consciousness, derives from a formative and semantic symmetrical structure that is circular and repetitive. The greater part of each of the three "layers" of this field is unconscious and implicit meaning, that is, composed of the circular, "enfolded" relations of symmetry. As a consequence, the smaller part of each of the three "layers" of this field of consciousness will be the explicit, visible, differential world of "unfolded" meanings. Consciousness, as a universal, infinite, and eternal field, is therefore a vast asynordinate system having the greater part of its structure always symmetrically enfolded in implicit and spiritual potentials.

The theory that is presented here is systematic and coherent and contains a detailed explanation of the organization and structure of consciousness. This is not to say that it dispels all the mysteries of consciousness. It does not. But neither do I accept that these mysteries are beyond conceptualization, or that the ineffable is foreign to science. I am optimistic that such mysteries can be accounted for in terms that link the effable with the ineffable within a field of study that can be called the science of consciousness.

Glossary

This book is written from an isomorphic perspective. Therefore, various terms are used in ways that differ from conventional usage. This glossary may help the reader to appreciate the isomorphic implications of these terms and the manner in which they have been used.

Asymmetry. Usually an irreversible relationship, but in the present theory a concept expanded to involve the set of relations that make up any system. Asymmetrical relations are a series of nonsymmetrical relations within the context of symmetrical relations. This is a system. As symmetrical relations are hidden from conscious awareness and nonsymmetrical relations never exist on their own, the smallest unit of conscious awareness is always a system of asymmetrical relations.

Asynordinate. A term taken from the work of David Bohm. It is used in this book to denote "degrees of implicitness" as well as the integration of the explicit with the implicit. As hidden implicit meaning is the foundation of, and makes up the greater part of, all systems, the smaller part of all systems is constructed from visible, explicit meaning. The integration of these two is an asynordinate relationship. It is like the relationship between a wave and the ocean. The two domains, the hidden (the ocean) and the visible (the wave), are never separated; rather, they are always integrated and therefore asynordinate. Since every system is asynordinate, discourse, subjectivity, and the physical universe are constructed from different forms of asynordinate systems.

Closure. A term taken from literary criticism, but used here to refer to the unconscious processes that construct units, unities, and totalities, the meaning of which is the basis for differentially knowing the world. In general a closure produces a difference through unification; it is a completion or conclusion brought about by the closing off of meaning from its contextual base. There are three kinds of closure: necessary, relative, and absolute.

Consciousness. Seen here as having three general "layers" or contexts: (1) a holistic, cosmic, and divine context from which all material existence emerges; (2) individual subjectivity, which also emerges from the divine cosmic context; and (3) discourse, which is the manifestation of cultural consciousness and which emerges from the context of subjectivity. These three contexts occupy a single continuous field of meaning, the foundational structural principle of which is symmetry; the basic function that links the three "layers" is recursive symmetry. The semantic structure of subjectivity represents the relations of symmetry, nonsymmetry, and asymmetry. Unconsciousness is therefore structured by symmetrical relations, and conscious awareness is structured by an asymmetrical combination of the relations of symmetry and nonsymmetry.

Discourse. A semiotic term, used here in the sense of an expressive practice, an active involvement of a person in communicative and semiotic activity such as reading, writing, listening to conversation or music, and so on. Mathematics is also a discourse. Discourse has an asynordinate structure and is seen here as manifest consciousness.

Environment. Always a problematic concept when we discuss perception. The problem is the impossibility of ever knowing the environment "in itself" as an unmediated physical reality, yet providing it with meaning within the context of the overall structure of subjectivity. From the present perspective, environmental objects are represented as *type 1/2* environmental systems—with the explicate order beyond our skins as a *type 1* system, our involuntary, prereflective transformations of them a *type 2* system, and the closure of their interaction a *type 1/2* environmental system.

While this is a cumbersome way of speaking, it does begin to suggest the kind of symmetrical complexity involved in what appears as simple, given "objects." This approach assumes that symmetry is the structural base of the physical world as well as of subjectivity.

Explicate order. A term coined by David Bohm to refer to the secondary order of the universe in which information is "unfolded" from the implicate order, which is primary. Bohm's term is used here as synonymous with explicit meaning. Explicit meaning is visible and differential and represents the structure and content of conscious awareness. Explicit meaning begins as hidden, implicit meaning that is then "unfolded" and made visible in conscious awareness.

Implicate order. Another concept taken from David Bohm. It refers to the primary order of the universe that "enfolds" meaning. It is synonymous with implicit meaning. Implicit meaning refers to that which is hidden or enfolded within the unity of a system. This book proposes that this character of "enfoldedness" or "hiddenness" is due to symmetrical relations that are the foundation of the three "layers" of consciousness.

Isomorphic. Pertaining to forms that are similar or the same. The prefix *iso-* is from the Greek *isos,* meaning "equal," while *morphic* pertains to "shape, or "having a form"; in combination they mean "corresponding to or having the same or similar form." An isomorphic paradigm or perspective represents a holistic way of viewing the world. The idea has been taken from the work of N. Katherine Hayles.

Language. A term applied to linguistic codes, often to falsely suggest that they exist as a formal entity independent of subjectivity.

Meaning of meaning. Used by Ogden and Richards in their book, *The Meaning of Meaning.* It refers to the basic and universal structure of meaning, that is, meaning before symbolic representation and interpretation. It is used in this book in the same sense.

Nonsymmetry. The relationship of difference. Nonsymmetry is a secondary relationship that emerges out of symmetry through a process of negation that comes from two symmetries interacting. The relational basis of distinctions, discriminations, sharpness, marking, and feature is nonsymmetry, and for the present theory these nonsymmetrical processes represent the structure and the content of conscious awareness.

Orientation. The primary value or focal point of meaning in a paradigm of consciousness. An orientation is a predisposition, but in relational terms it means the central, recursive values of the paradigm.

Paradigm of consciousness. The organizational framework of knowledge, meaning, and behavior. This sense is distinct from Thomas Kuhn's notion of a paradigm of method and application. There are only three paradigms of consciousness and so only three ways in which knowledge, meaning, and behavior can be organized. These are a symmetricized paradigm, a paradigm of reason and an isomorphic paradigm. *A symmetricized paradigm* orders meanings in a traditional irrational manner, so that it forms symmetry sets. Everything within the set equals everything else: A equals B equals C and so on. This ordering is typical of body sensations as well as the repressed unconscious. *A paradigm of reason* orders meaning in a linear, mechanical, and/or reductive manner. This paradigm has a predisposition for making differences exclusive, fragmenting knowledge and behavior. *An isomorphic paradigm* orders meaning in a holistic, nonlinear manner and is disposed to integrate differences into larger or smaller contexts.

Recursive. A term from chaos theory. In this book it refers to circular and repetitive processes. A process can be recursive in respect to the three relations of consciousness. Hence, recursive symmetry will produce a symmetricized perspective, recursive nonsymmetry will produce a paradigm of reason, while recursive similarities (systems) will produce an isomorphic paradigm.

Signified. In the linguistics of Ferdinand de Saussure, a concept and/or image. In this book, the term is used in the same manner. In linguistics, signifieds are generated by the selection and combination of signifiers into words, then into sentences, and then into texts.

Signifier. One aspect of Saussure's sign structure (the other being *signified*). As used here, it refers to the second level of the isomorphic sign. From the present perspective signifiers are made up only of visible, discrete keystrokes such as letters of the alphabet, numbers in mathematics, or notes in music.

Subjectivity. The second general context or "layer" of consciousness, that is, individual consciousness. In combination with the first "layer," cosmic consciousness, and the third, discourse, it forms a unified and continuous field of meaning and consciousness. Each "layer" is structured by the relations of symmetry, nonsymmetry, and asymmetry. The semantic processes that structure individual subjectivity also operate within the cosmic context.

Symmetry. In its common meaning, "sameness." In this book, symmetry is the structural, immanent, formative, and fundamental relation that makes up the hidden, unconscious background context of consciousness.

Symmetry transformation. In mathematics, any change within a system that does not destroy the coherence (identity) of the system. For the present theory, a symmetry transformation makes conscious that which has been unconsciously hidden within the symmetry of the system. The concept implies that every system has the potential for transformation. Any constituent can be a system, and so a symmetry transformation can change a constituent into a context of a smaller system or can transform a context into a constituent of a larger system.

System in general. The whole of subjectivity as a system. In describing the relationship between subjectivity and the environment, a term is needed that avoids implications of "objectivity" or solipsism. I have used *system in general* as this term.

System type. A second term used to describe the relationship between subjectivity and the environment. It refers to the structural and determining effects of the environment in the processes of constructing perceptions or sensations.

Part 1

The Holistic Context

Chapter 1

Approaches and Methods

The question "What is consciousness?" has become increasingly important to Western science in the last quarter of this century. For example, this was a central question for the conference "Towards a Science of Consciousness," held in Tucson, Arizona, in 1996. Debate at that conference appeared to flow in and around the proposition that the "what is" question was the hard problem. The easy problems apparently are those currently being tackled in neuroscience laboratories around the world. These sorts of problems deal with such things as sensory coding and decoding, the link between the artificial intelligence of machine minds and our minds, information processing, and so on.

With no Western theory of consciousness there are a range of possible responses to the hard question. On April 16, 1996, the *New York Times* reported that the Tucson conference delegates had four responses to the problem of what consciousness is. These were that "it does not exist, it will be answered soon by conventional science, there must be something else in the universe that we do not understand and hey, guys, forget it, we can never understand consciousness." Even if we disregard the simplicity of a newspaper report, this "landmark" Tucson conference must be seen as a general acknowledgment that conventional scientific methods have failed to produce a theory of consciousness.

Is that failure endemic to "classical" scientific method, or is it due only to a lack of research? I suggest the former is the case. R. L. Amoroso, one of the delegates to the Tucson conference, suggested, "In the history of scientific thought problems have been typically hard only when the framework for posing the question is not well understood." ("Noetic Field Theory: the hard problem becomes easy," paper, Tucson Conference, 1996.) The framework for the "what is" question has historically come from within the domain of cognitive science and been framed by the sight lines of rational materialism. While the protocols of these sight lines are generally well recognized and strictly policed in cognitive science, in discursive practices the hidden assumptions, predispositions, rhetorical influences, and reinforcements are less well understood.

Mechanical and reductive methods and discourses are useful for solving puzzles within the conventions of rationalist and empirical sciences, but such approaches fail with the question of consciousness. They fail because rationalist and empirical methods inherently work to clarify by isolating and excluding material from its broader contexts. These exclusions and erasures operate against the grain of consciousness, which, I will argue, is an inclusive, circular, and holistic phenomenon. Like quantum physics, consciousness cannot be analyzed and treated by reductive or mechanical means; therefore, cognitive science, with its present rational, materialistic approach, is unlikely to find the answer to the question, "What is consciousness"?

Any description of consciousness is itself a circular act in that it attempts to reflect on reflection, describe description, observe observation, and be conscious of the processes of consciousness. This circularity appears to be an essential characteristic of consciousness and also of the discursive processes of describing and analyzing consciousness. It needs to be noted that circular and holistic methods of analysis are different from the mechanical and reductive methods traditional in the rational and empirical sciences. The key feature of a holistic approach is the manner in which information and meaning is organized.

The organizational principles of a holistic approach are the principles of recursive circularity and inclusion. The principle of inclusion states that all distinctions, differences, and parts are treated, not exclusively, but always as inclusive details or constituent features of larger wholes or contexts. No distinction or part is seen to be so different as to negate its connection with its contextual whole, and therefore no distinction simply carries one exclusive and limited meaning. Distinct features and differences and also parts are always linked together into larger wholes through their underlying similarities. In this manner, a part reflects the whole, and the whole will mirror its parts. Also in this manner, constituents will always have a context. Yet the dialectical relationship between constituents and context is not simply a general and static circularity. Rather, it has recursive potential. This means that all constituents can themselves become contexts for "smaller," internal constituents, and this reductive process goes on until the smallest context transforms into the largest, universal context—often referred to as God. In like manner, contexts can become constituents for larger contexts, and this transformation can recur many times over until the largest context—God—is reflected in the infinity of universal constituents. This holistic and quite universal constituent/context relationship is significant to the present endeavor to define the question of what consciousness is.

These holistic organizational principles contrast with mechanical and reductive methods of organization. The key feature to a mechanical or reductive method (inherent to a rationalist/empirical approach) is, again, the principle of exclusion. All information and meaning is organized by applying the rule of exclusion to distinctions. Differences are thus treated as if they could function as separate and exclusive states and as such could give rise to such pursuits as Beingness, the concept of separate identities, binary logic, dualistic thinking, the separation of mind and brain, decontextualization of information, the instrumentalization of terms, a Platonic view of language and mathematics, and a preference for the formalism of differential grids, networks, and calculus. This differential organization of meaning has led inevitably to the fragmentation of knowledge into distinct academic disciplines with their associated specializations and experts. In a more general sense this differential organization has produced the separation of knowledge and knower, observer and observed, emotion and intellect, madness and sanity, mind and body, and so on.

Because consciousness is constituted by circular processes, in any analysis it should be treated by an appropriately circular, holistic method. The circularity of consciousness is not, however, something we can easily observe, like a frog under a microscope. It is hidden and embedded in every description of consciousness, no matter how incidental. This circularity is contained in the set of tacit assumptions that underpins every approach to consciousness. For example, a rationalist approach to consciousness posits the self-evidence of reason as the key moral, and sometimes divine, feature of consciousness. An empirical approach assumes the primacy of sense impressions, while a rationalist/empirical approach assumes the existence of a Cartesian cogito. Cognitive scientists are prone to assume the ubiquitous presence of this cogito (which is then never interrogated) in the process of focusing on the minutiae of physiological features, information processing, or sensory coding, or one of the many other surface questions common to cognitive science.

Yet while the investigative processes of cognitive science cannot escape the circularity inherent in discussing consciousness, neither does the present holistic approach elude this circular embeddedness. Thus a holistic view of consciousness will always presuppose the inclusion and integration of an observer and the observed, and all that that implies about the oneness of knowledge, knower, and known.

The first questions to be asked when tackling the question of consciousness must therefore be "What is the approach?" and "Are the approach and methods appropriate to the conditions of consciousness?" It should be stated here that the approach used in this book is explicitly holistic and inclusive, and the non-Cartesian ramifications of this perspective will become progressively clearer.

Domains

An additional concern for any study of consciousness is the question of domain—what is the domain of consciousness? Domains can be limited and exclusive or universal and inclusive. For example, the rational materialist view limits the domain of consciousness to "the subjective" and often to the introspective area of "mind." Such a domain as the mind is currently taken as having internal mental content or qualia that are directly knowable to the conscious person in acts of introspection. This exclusive internal domain of the mind "in here" is seen as separate and different from the other exclusive external world of the environment "out there." In cognitive science almost all debate takes place within a Cartesian domain of the rational, subjective, and introspective mind. Some of the effects of using this exclusive domain are that cognitive science will favor cognition over consciousness, information over meaning, coding over discourse, computations but not unconsciousness, reason but not madness, materiality but not illusion, observations and physical processes but not philosophical theories, facts but not God.

In contrast, an inclusive, holistic approach to consciousness must have a universal domain and so exclude nothing. Each one of the above-mentioned pairs will be part of any discussion of consciousness and therefore must come within the boundaries of an inclusive approach. For example, an inclusive, holistic approach to consciousness has to be able to integrate all the features of consciousness, such as the universals of love and faith as well as madness and reason and the more formalistic manifestations of computations, calculus, and codings. It must have something to say about our relationship to God. In a similar manner both the implicit and the explicit will have a significant place within a universal and inclusive domain of consciousness, as do such notions as reality, illusion, and hallucination.

By its very nature an inclusive domain has a universal range of application, for nothing stands beyond or outside it. A universal and inclusive domain of consciousness means that consciousness is an aspect of everything—everything physical, biological, or even formalistic—in the universe. Such is the circular logic of a universal domain that incorporates an inclusive and holistic approach to consciousness. The universal domain may appear too unwieldy for the tidy minds of most cognitive scientists, but this breadth is entirely necessary to analyze and encompass the totality and holistic applicability of consciousness.

Philosophers like Schelling, Hegel, and Spinoza were concerned with the universality of consciousness, as are many religions such as Hinduism, Buddhism, and Taoism. The universal domain is also the domain employed by the Catholic mystic Teilhard de Chardin, who saw Christ in "matter" and God in evolution. He held that the universe is ultimately spiritual in nature and slowly developing toward greater perfection and unity. The Protestant theologian John Cobb appropriates the universal domain of consciousness within his interpretations of "process theology."[1] A universal and holistic domain is also common to Eastern mysticism. Fritjof Capra points out in *The Tao of Physics* that Eastern philosophy has constantly referred to an "ultimate, indivisible reality which manifests itself in all things, and of which all things are part."[2] These comments echo the inclusive approach taken to quantum mechanics by David Bohm and Basil Hiley in *The Undivided Universe,* in which they assert that "there is no real difference between mind and matter, psyche and soma [and that] a rudimentary mind-like quality is present even at the level of particle physics."[3]

It is evident that the use of inclusive, holistic methods and universal domains are common to a variety of work concerned with consciousness outside the fields of cognitive science; it is this inclusive and holistic approach I will be using to define consciousness within the universal domain.

Relations

The project of this book is to produce a systematic and coherent theory for the question "What is consciousness?"—a state that has hitherto defied explanation in Western terms. How to do this? Given the holistic and universal breadth of our approach, it seems important to use terms that appear to have total and universal application. I have already implied that the present approach assumes that consciousness has a universal application (that nothing stands outside it), but we cannot develop an explanation very far by continuing to just use the term *consciousness.* We need a set of systemic terms equivalent to *consciousness* that can articulate, in holistic ways, some of the circular processes and structures of consciousness.

One popular candidate in the area of cognitive science is the term *information.* If we assume that the content of consciousness is information, then the inclusive question to be asked about information is not about how minds and computers use it, but "What is the context in which information occurs?" Information can be numbers, codes, written communication, facial expressions—almost anything can count as information. Genetic codes as well as political speeches impart information, and it is this magnitude that suggests the possibility of a universal application.

Yet the term, because it is often used as if it were devoid of any context, carries a surplus of baggage and so implies too much to be very helpful. The notion of information is like a black box, full of hidden meaning that is never questioned, but that is nevertheless employed in building instrumental systems of explanation. *Information* is like *cognition* or *sense data* or a range of other instrumental terms that explain nothing in particular but simply act as broad linguistic pegs from which a whole range of arguments are conventionally hung.

The term *information* always implies consciousness for it cannot be created or produced without consciousness. This is not so much an issue about whether lower life forms have a consciousness or not, but about the instrumental manner in which the term information is often used. To challenge this use we need to focus on context. The context in which information occurs is within communication. To condense a great deal of contemporary debate, we can say that all information can be seen in terms of discourse and signs that carry meaning. From the philosophy of language we know that signs are not the single-dimensional entities that are often referred to by mathematicians and computer experts. Rather, a sign has several "vertical" levels that function simultaneously within its "horizontal," linear dimension. Signs are not so much single ontological entities as small complexities of constructed relationships, like a "packet" of meaning created within the dynamic processes of a discourse. The term information, used in communication context, expands into a sign, and if signs are carriers of meaning and meaning is concerned with relations, then here are three terms more appropriate and more basic to the processes of consciousness than *information.*

But the terms *signs, meaning,* and *relations* appear vaguer than the hard edge of "information" that can be computed in the form of analog or digital data. The significance of the difference between these "vague" terms and the precision of *information* or *data* or *facts* is that the vague terms are inclusive and the precise ones are exclusive. Their precision is obtained through the exclusion of the systemic context in which they occur. The exclusion of these contexts puts in doubt the instrumental meaning inherent in *information.* Therefore, in order to conform to an inclusive approach to consciousness it will be necessary to use a set of basic terms that also are also inclusive and to avoid using, as much as possible, terms that are exclusive. For these reasons I will not be using the term information (or any of the other instrumental terms from cognitive science or behaviorism) to describe consciousness but will rely on the vaguer but more inclusive terms of relations, meaning, and signs.

Relations are universal. Gregory Bateson has argued that relations should be used as a basis of *all* definitions.[4] If this happened, it would produce a paradigmatic shift from a focus on "substance" and differential objects to a focus on the integrating relations of their social contexts. One aspect of this shift is a move from the discourses of dualism to those of nondualist integration and holism. Dualism develops and is sustained by a focus on differential objects and by a set of materialistic terms that imply substance, such as *properties, property, substance, mass, assets, holdings, ownership,* and so on. Charles Birch suggest the term "substance thinking"[5] for that kind of materialistic focus common to the mechanics of science and, I would say, much of economics. Substance thinking, with its dualist strategies, is impossible, however, if we use discourses that describe through relations. A focus on relations always produces an integrated, holistic perspective. As the present approach is holistic, we will begin with an explicit focus on relations.

We commonly use relations in describing love, society, the environment, the processes of quantum physics, matter, energy, madness, the cosmos, and concepts of God. Relations structure mathematics, discourse, and signs, as well as all nonverbal language, including sensations and feelings. Relations give coherence to every system; they provide the structure and content of concepts; they produce our sense of order and disorder as well as creating meaning. Every law of physics is concerned with relations, every system of government is about social relations, and every group represents a combination of interpersonal and cultural relations. As there is nothing beyond the boundaries of the totalizing framework of relationships we can therefore hypothesize, from a holistic perspective, that the structure of consciousness is constructed from relations. The term *relations* is thus a prime candidate for describing consciousness and its contents, and its universal and radical nature fits well with the holistic and integrating approach.

Modern science finds that relations permeate every facet of matter, motion, and life. For example, molecules are seen as the building blocks of life—they are the basic substance out of which everything we see, hear, smell, touch, or taste is made. But if you divide a molecule into its constituent atoms, the relations that construct it and maintain its existence as an entity transform. Under these transformed conditions it will cease to have chemical properties but takes on other, unstable properties. Yet the constituents of the molecule—atoms—are themselves further contexts for yet other constituents. For example, an atom is made up of a tiny nucleus and a cloud of electrons that buzz around it. The relationship electrons have to their nucleus—how many electrons and their orbits about the nucleus—determine the interactions between atoms and hence the laws of chemical combination for the formation of molecules.

But the story of relations does not end there, for the tiny constituent nucleus, containing most of the mass of an atom, is the context for its own constituents, two major particles called protons and neutrons. And these constituent particles are in turn not the final building blocks for matter but represent contexts for further constituents—the meson that transforms into yet a further context for more constituents and so on. In each one of these transformations, from molecule to atom to nucleus to protons and neutrons to mesons et cetera, there is a transformation of relations and a change of energy. If relations have energy then it explains this coincidence of transformed relations with changes of energy at every level of the material world.

But if the term *relations* can be used as one of the basic terms to describe the processes of consciousness, what are some of the implications this word contains? My understanding of relations is that they are often thought of as "objective" or formal properties that belong to physical bodies, like a connecting signal. For example, the term *interactions* is often used in physics to refer to a range of relationships that particles have to each other. Interactions and relationships are seen here as identical. Yet the term *interaction* tends to imply some kind of physical connection or property belonging to particles. But this is not the case, for the interactions of particles represent the social relationships of the group, which are the property of no single particle. To describe interactions as the social phenomenon of a group of particles is to shift the meaning of interactions as a sometimes passive, possessive property belonging to a particular physical entity (a particle), to the concept of an active social agency in its own right, having its own power and energy that affects and creates group behavior.

Relationships always pertain to the interaction of groups, whether the group is one of particles or people. In this sense relations construct groups. Thus relations are never the property of a particular individual entity, but neither are they the simple mathematical sum of a group's interactions. If, however, we were content with this latter mathematical conclusion, then we would be assuming that relations are the passive property of groups alone rather than the passive property of individuals. What is wrong with this usual conclusion?

The difficulty with this instrumental view is that in its dismissal of relations as only "passive properties" it never asks how groups are constructed or what constitutes relations. Because these questions remain tacit, relations are automatically denied a constitution beyond being an attribute, and so the animate power that has constructed the group is ignored. Yet if we ask what constitutes relations or how relations construct a group, then relations are quickly transformed from passive attributes into active and animate agents. As active agents, relations become self-organizing and are seen to have social, group, or field potentials. Such potentials will involve the power and energy to create group behavior in particles and people, which is just another way of looking at an energy field. As all particles and people always operate as group members, the relations that have created this group or field and are not attributes of any one individual must have transpersonal potentials. Relations therefore appear to represent a field dimension of energy and power that is prior to individual physical entities such as particles or people.

As long ago as the end of last century, science began to view energy fields rather than matter as primary. And instead of trying to explain fields in terms of matter, physicists began to explain matter in terms of fields of energy. Somewhat similar debates took place in the humanities, where nineteenth-century liberalism saw the autonomous individual as primary while Marxists placed the social field as primary. Liberals and Marxists have tended to construct a dualism between the individual and society that is unsustainable when the focus is on social relations. A dualist division between energy fields and matter is also unsustainable when particle interactions are the focus of attention. Dualism is always negated by a focus on relations and always created by a focus on entities.

From a holistic perspective, the context of a field, for both particles and people, will always have an emergent and symmetrical relationship to the various constituents of that field. Relationships provide the essential structure of background fields as well as being the ultimate formative cause for the realization of individuals within those fields.

Western science has discovered a variety of different fields that produce a range of different interactions or relationships. The four common fields of interactions are gravitational interactions, electromagnetic interactions, the weak interactions, and the strong interactions. In addition, there are gauge fields. From gauge field theory we learn that gauge fields are supposed to give rise to the four fundamental interactions in the universe. According to unified-field theory these four fields are the asymmetrical remnants of a once perfect world of symmetry at the time of the big bang. Symmetry and asymmetry are, of course, relations, relations that, according to these theories of contemporary science, have the potential to create the physical mass of the known universe and to bring order to it through these four fundamental fields of interaction. Such is the constitution and formative power of relations.

From a metaphysical view this dimension of energy and power is represented by a cosmic intelligence or God. Does God or cosmic consciousness come in the form of relations? If he does, then he would have at least two of the attributes implied by the above discussion: omnipresence, being in every part of the universe for all eternity, and omnipotence, being able to create the physical universe.

Another characteristic of relations, referred to in passing above, is energy. An energy with form is the description used by David Bohm and F. David Peat for the idea of "active information." Speaking about quantum theory, they say, "The basic idea of *active* information is that a *form,* having very little energy, enters into and directs a much larger energy. This notion of an original energy form acting to 'inform,' or put form into a much larger energy has significant applications in many areas beyond quantum theory."[6]

As I have stated earlier, the term *information,* whether active or potentially active, is problematic for a holistic approach to these questions. More appropriate is the term *meaning,* which Bohm himself uses in another context, and the term *relation* that is proposed here. By using these terms instead of *information,* we can say that the active and self-organizing relations (as energy forms) inhere in and organize everything in the universe, from the smallest quantum matter fields to the movement of galaxies or the interactions of human society.

But what kind of energy do relations possess? The concept of active relations has similarities with the notion of vibrations. We know that vibrations are as omnipresent as relations; that at every level of organization of an organism there is evidence of rhythmic oscillations, vibrations, periodic movement, or cycles. In all atoms, molecules, and cells, vibrations occur all the time. Are vibrations active relations? They certainly appear to be the same. For relations to be active there has to be activity of some kind. The most immediate kind of surface activity is vibration. If this is the case, then all relations will have resonance, and different relations will produce different kinds of resonance. Resonance, therefore, can be thought of one of the active characteristics of relations.

Yet another characteristic is meaning. David Bohm proposed a link between meaning, matter, and energy. Bohm suggested that the three are reciprocal; that "energy enfolds matter and meaning, while matter enfolds energy and meaning."[7] (Bohm was using the terms "enfold" and "unfold" in their commonsense use of folding inward and folding outward). In this century Albert Einstein has made the symmetrical relationship between matter and energy a commonplace in Western science, but the connection between energy and meaning is novel.

If we accept Bohm's proposition that energy enfolds meaning as well as matter, then the processes of enfolding and unfolding implicitly involve the transformation of relations. In every transformation it is relations that are transformed along with a variety of physical properties and changes in mass and energy. Even in mathematical discourses transformations of mass, energy, relations, and meaning occur within the mathematicians involved in that discourse. Whatever else happens in a transformation, relations are always involved. And so as energy changes its form into matter, so the relations inherent to this change are transformed.

To support Bohm's contention that meaning is also involved in these transformative processes I foreshadow here a later discussion that proposes that the structural character of meaning is always provided by the relations that constitute it. In essence this means that relations are always relations of meaning and that meaning always has a structural content constituted by its relations. This being the case, Bohm's proposed equivalence between meaning, energy, and matter is supported here because any change between matter and energy will require the transformation of relations and all relations are relations of meaning.

If all relations are always relations of meaning, then meaning will be a universal and omnipresent feature of the entire universe, at every level, in every shape, in every form, and in all organizations. This totalizing sense of meaning can be seen as roughly representative of the concept of omniscience—as being all-knowing or having infinite knowledge. Omniscience is a theological term that I do not want to pursue here except to say that the kind of knowing referred to in this theory of consciousness is never a Cartesian form of explicit awareness. The term *omniscience* is therefore used here in a non-Cartesian and spiritual sense that will become clearer as we progress.[8]

The synthesis of energy, matter, and meaning proposed by Bohm and added to here could appear at first to give support to the old, much-maligned metaphysics of vitalism. Vitalism is a doctrine advocating a life-giving energy or substance usually seen as independent from matter. Since Aristotle, various forms of vitalism have been used to account for biological phenomena. Modern biologists condemn them all as pseudoexplanations, holding that life is an emergent property due to new organizations and the transformation of the relations between matter and energy. Given that it is the symmetrical connection between matter, energy, and meaning that Bohm was concerned with and that his contention is supported by our discussion on relations, one wonders where the modern biologist will turn to disavow these present proposals, which may or may not lend aid to the ancient faith of vitalism but which, in any case, support the general proposition of an animate universe.

As for metaphysics, any proposal that focuses on relations, ipso facto provides a metaphysical perspective. Relations are never material, even when they are postmaterial, prior to material existence; they are always metaphysical phenomena. It is interesting to note that Bohm's proposal of incorporating meaning into energy and matter has some parallels with pre-Cartesian thought as well as with Teilhard de Chardin's panentheistic notion of cosmic evolution in which everything in the universe is seen to have both a physical and a spiritual aspect.

Beyond accounts of Western biology and the question of evolution, the linking of meaning with energy and matter reflects a synthesis found in Indian metaphysical philosophy and the spiritual notion of cosmic intelligence. In *The Concise Yoga Vasistha,* a leading scriptural text on Hinduism, we are told that "the world appears to be material, yet in reality it is pure consciousness."[9] And further, "infinite consciousness is present everywhere equally. The differences are only due to the intelligence identifying itself as different substances. . . . The mountains, the forests, the earth and the sky—all these are but infinite consciousness."[10] If we think of relations as energy and meaning, then the inclusive logic of this will lead us holistically to the notion of a cosmic intelligence.

This holistic conclusion is reached by the following path. Energy can enfold meaning (relations) as well as matter. Meaning is thus enfolded within all matter in the form of relations, and these relations have been created by energy, which is in the form of relations. In this sense relations represent the omnipresent, omnipotent, and omniscient potentials of a cosmic order or consciousness. In this universal sense, relations are essentially made of the same "stuff" whether they are relations at a quantum level of particle interaction or at a social level of community or as a feature of subjectivity. While there are three types of relations (discussed in the next two chapters), they do not differ as category types throughout the universe. All relations will have the same qualities of energy, resonance, meaning, and consciousness. These common qualities mean that subjectivity, which is created from relations, is able to perceive and reflect on the universe in all its myriad forms because it is made from the same relations as the patterns that are perceived. In this sense subjectivity, as a symmetrical aspect of the divine, is simply reflecting on itself when perceiving the world.

For my own part I take it that relations are always relations of meaning that have energy and ubiquitous, formative potentials. As active, self-organizing agents, relations will provide the force and the structure for the transformations that unfold and enfold matter. Thus, through this self-organizing potential, matter can be created and destroyed by the relations that constitute it. A holistic approach to relations will therefore view them as the evidence of a cosmic intelligence, which means that the subjectivities of individuals will be symmetrically reflected in a cosmic intelligence and a cosmic intelligence will be symmetrically linked to the subjectivities of individuals, which are but constituents of this universal context.

In a more mundane sense, a system of relations can also be thought of as a field or stream of subtle, meaningful energy. Examples of a stream of meaningful energy are discourse or, say, mathematics. These are complexes of interrelated systems of meaning and can readily be thought of as containing an energy or force. Discursive force is clearly apparent in successful advertising and political rhetoric. Mathematical force is always evident in the "beautiful mathematics" of symmetry.

Relations in this holistic sense, then, are not simply the passive attributes of active physical or formal entities, or even the discovered and passive links of a physical and biological world but are themselves active, resonant, meaningful, and self-organizing agents that transform themselves and so change states of energy and mass in the realization of the material world. As such, relations constitute the omnipresent, omnipotent, and omniscient evidence of an infinite cosmic consciousness. This conclusion is similar to the Hindu belief in the self-organization of cosmic consciousness, whereby "infinite consciousness is present everywhere equally and where differences are only due to the intelligence identifying itself as different substances."

Summary

This chapter introduces some of the implications contained in different approaches to questions about the nature of consciousness. Cognitive science is prone to use reductive and mechanical approaches to consciousness that narrow and reduce it to the Cartesian domain of an individual mind. Contrasted to this approach is the present holistic analysis, which proposes that the appropriate domain for an inquiry into consciousness is the universal context in which a cosmic consciousness holistically reflects the attributes of the subjectivity of individuals.

Some basic terms that could be used to describe consciousness are discussed. The term *relation* has universal application and is therefore seen to be an appropriate descriptive term for a universal approach. It is argued that relations are not the passive attributes of a physical world or even the property of a discipline like mathematics, but rather are themselves active, organizing, and meaningful energy forms that can transform states of matter and produce the realization of the physical world. As such *relations* is the ideal descriptive term for an inquiry into consciousness.

Chapter 2

Types of Relations

In the previous chapter it was proposed that relations are the universal, active agents of cosmic and individual consciousness. It was also suggested that relations have an active, self-organizing intelligence that gives rise to the physical forms of nature, and that relations provide the structure of consciousness and thereby a means for discussing the content of individual consciousness. In this chapter I want to say something about types of relations and to look specifically at the relation of symmetry.

Symmetry

David Peat tells us that shortly before his death Werner Heisenberg, the creator of quantum theory, argued that "what was truly fundamental in nature was not the particles themselves but the *symmetries* that lay beyond them."[1] (original emphasis) Peat suggests that these fundamental symmetries can be thought of as the basic ground of all material existence. As such they can be seen as representing an ultimate level of reality. If this is the case then it would mean that the elementary particles of quantum physics are the material realization of these fundamental symmetries. In other words, such particles have emerged from these symmetries through a series of complex transformations. As manifestations of this fundamental metaphysical level, electrons, mesons and protons, the building blocks of physics, would therefore represent a secondary level of reality rather than the first as held by materialists.

A symmetry of course is a relationship and a complex of symmetries is a complex of symmetrical relationships or a field of symmetry potentials. Symmetries are difficult relationships to come to terms with, both because of the weight of meaning they carry concerning consciousness and the physical world, but also because the relationship of symmetry is hard to conceptualize. I return to the problems of conceptualization later, but for now let us focus on some of the inherent characteristics of symmetries. Peat proposed that the character of this metaphysical level of reality is constitutive and descriptive, but in addition, symmetries "have an immanent and *formative* role that is responsible for the exterior forms of nature."[2] (original emphasis) Here we have four features identified as functions of symmetries. They are: constitutive, descriptive, immanent and formative and they begin to form a description of the nature of these universal relationships.

I interpret these four functions to mean that symmetries are a constitutive feature of everything, that they are immanent in everything and that their formative role comes from a self-organizing and universal energy that gives them the power to materialize the physical forms of nature. Symmetries are therefore not passive relations that are innocently discovered in the physical world or society by objective observers. Rather they are formative, and thus omnipotent agents operating at every level of the universe (omnipresent). Peat's descriptive role is used in a passive sense of "symmetries are a way of describing complex phenomena."In contrast I would propose that the descriptive role of symmetries is also active, in that all symbols, signs, discourse and communication are constructed and produced through symmetrical processes. (Symbolic processes will be discussed fully in chapters 8, 9 and 10.) In addition to these four functions nominated by Peat, I suggest that symmetries also have semantic structure and so are able to play an ordering and meaningful role in every aspect of the universe. This last characteristic refers to the non-Cartesian meaning of omniscience.

Omnipotence, omniscience and omnipresence are characteristics often attributed to God. It is proposed here that symmetry and its potentials are isomorphic with the concept of God as a cosmic consciousness. The relations of symmetry are also a productive way of referring to the structural content of individual and cosmic consciousness as well as the self-organizing and universal energy immanent throughout the physical universe. In other words, the relations of symmetry is a fruitful way of discussing and referring to God-like actions, structures and effects. In this present endeavor I am interested in developing a Western discourse based upon the circular logic of symmetrical relations rather than using the discourse of traditional mysticism, religion or spirituality, while in no way denying these. It should be recognized however, that all discourses are inherently limited by the formal symbolism from which they are created. In other words, discourse is not a path by which one develops faith, discourse simply provides a traditional and efficient means of communicating with others.

Drawing on this discursive focus we see, from the discussion above, that symmetries have a sixfold purpose: (1) descriptive (or symbolic), (2) constitutive, (3) immanent, (4) formative, (5) structural and (6) semantic.

If we take seriously Heisenberg's belief in the fundamental nature of symmetries, then the basic ground of all material existence is the constitutive and formative fields created by symmetrical relationships. In other words, the formative power of symmetrical relationships has produced material existence through the symmetrical fields from which such physical entities arise. This is saying not much more than we find in orthodox physics, which theorizes that the mass of particles is created by their interactions with other particles and that particles are the manifestations of the underlying reality of their fields. From the present perspective these underlying fields have a symmetrical structure, being composed of symmetrical relations that have formative and semantic potentials. This formative energy, (having meaning and consciousness) of fields will therefore be immanent in all material existence. From this perspective the differences between the inorganic and organic is found in levels of complexity of organization and therefore in consciousness. The difference is not in the categorization that only the higher forms of life have consciousness while the rest of the inorganic universe misses out.

A similar view about underlying metaphysical structures was propounded by C. G. Jung. Jung, in common with many other thinkers throughout Western history, believed in an ultimate unity of all existence. This unity is structural yet is outside the categories of time and space and beyond our separation of the physical and the mental. Jung referred to this unity in medieval terms as *unus mundus*. Jung saw that the idea of *unus mundus* was founded on "the assumption that the multiplicity of the empirical world rests on an underlying unity."[3] The divine background for our emergent, empirical world of differences is thus the *unus mundus*.

Peat did not recognize the link between symmetry, meaning and consciousness but, no doubt influenced by Jung, does goes so far as to ask, "Is it possible that archetypal symmetries of this nature could also manifest themselves in the internal structure of the mind?"[4] The rhetoric of this question looks to the propositions argued for here, which are that complex symmetrical fields of relationships are the fundamental ground of all material existence but also the fundamental structure of consciousness, both in its immanent and formative forms within nature as cosmic consciousness, and in its constitutive, structural and semantic forms as the content of individual subjectivity. Symmetries are therefore both the metaphysical, formative cause of material existence as well as the structural content of cosmic and individual consciousness. These conclusions find some resonance with the concept of "nonlocality."

Nonlocality is a concept that has developed out of the science of quantum mechanics. It refers to the symmetrical link between two spatially separate systems and it indicates that they are "together yet apart." For example, a subsystem or particle A can be spatially well separated from its companion B, yet the measurements of A will be not only depend on the results of B but also on what the scientist decides to measure at B. This "spooky action-at-a-distance" transgresses the certainties we have about a physical world based on the mechanics of cause and effect, at the center of which is the notion of locality. Locality implies that effects that are strong within a given region fall off outside that region and that the further the distance from the cause the weaker the effect. Locality is a central faith of Newtonian mechanics. The notion of locality helps divide the world up into its separate, self-contained regions and systems that then interact by forces and signals.

In the face of this well-established Newtonian view, nonlocality represents one of the most radical features of the science of quantum phenomena. From its implications comes the concept of wholeness, of an interconnected universe that is not partitioned off into separate regions but is totally interconnected. Every part of this interconnected universe will be instantaneously connected to the whole, and these relationships cannot be explained in terms of pushes and pulls or the interactions of physical signals. The wholeness of the quantum world is derived from what Bohm has controversially called the "quantum potential." It is controversial in the field of physics, for such terminology moves away from the old certainties of a Newtonian worldview and also away from the formalist belief in the completeness of abstract results derived from mathematical equations.

The concept of a quantum potential involves instantaneous connection rather than the expected retarded connection across a region of space. An instantaneous connection is the concept of symmetry or a symmetry connection. In the same way, an interconnected universe is a reference to the symmetry of the universe. Both interconnections and instantaneous connections refer to the nonspatial manner in which symmetry operates. Thus the concept of an instantaneous connection across space defies the common differential framework of a signal moving at the speed of light. A signal moving at the speed of light involves distance, locality, cause and effect differential space—all the discursive ingredients of an explicit and mechanical world. A quantum potential is a name for a set of scientific characteristics that are not of this mechanical world but relate to the hidden, implicate world of symmetry. It is no wonder Bohm's terminology is controversial in some quarters

A focus on relations will force us to ask, "What relations are involved in a quantum potential?" The answer falls out of the question, for clearly quantum potential refers to symmetry potential. Symmetry is the relationship in all non-local connections. Symmetry is the relation we refer to when speaking about "instantaneous" connections. Symmetry is the relation from which "wholeness" comes and by which wholeness and interconnections are built. The "spooky action-at-a-distance" is the effect of symmetry potentials and symmetry relations. Such effects involve a consciousness or a "knowing" by particles of each others, actions across these distances. How is this possible?

I suggest that it is not possible if we bring to this question a Newtonian view of a deanimated, mechanical world. Such a view will focus on explicit differences in which locality and cause and effect will figure prominently. Such a view will refer to constituents without too much regard for their contexts. Such a view will look to the particles as the object of "knowing" rather than to the symmetrical relationship between them and the experimenting scientist as the "knowing."

In contrast, a holistic view in which the universe is seen as animate will give priority to the quantum potential over the constituent particles or subsystems. Such a view will see consciousness or the "knowing" in the relations of symmetry, for all relations are relations of meaning and therefore the meaning of fundamental symmetries will be the kind of universal knowing that is not restricted to locality (an individual's mind for example) and the differential mechanics of cause and effect. Such an animate knowing will be universal, and its structure will interlink everything within the universe so that cosmic consciousness pervades the entire universe. This is the animate end point of logic in Bohm's "quantum potential."

A universe that is permanently and instantaneously connected at every point is a universe based upon the omnipresence and omniscience of symmetry. Such a universe is a place for the symmetry of an "active" cosmic consciousness. The evidence of nonlocality at a quantum level therefore points to an animate universe, to a cosmic consciousness, structured by symmetry relationships. In a perverse way, the smallest systems of scientific investigation have thus provided evidence of the largest context for all systems. This is a transformation of cosmic proportions.

Heisenberg did not say that symmetries are energetic relationships of meaning and as such represent the constitutive and immanent features of cosmic and individual consciousness. Such a statement would have meant going beyond the present confines of physics. Rather, symmetries seem to have been understood by Heisenberg, as well as many modern scientists, in Platonic terms as a priori, ideal forms existing within the physical universe but beyond individual minds. Dualist thinking dominates most contentious issues, and this one is no exception. In a discussion about the relative merits of the ancient philosophies of the materialist Democritus and the idealist Plato, Heisenberg comes down on Plato's side: "For the smallest units of matter are, in fact, not physical objects in the ordinary sense of the word; they are forms, structures or—in Plato's sense—Ideas, which can be unambiguously spoken of only in the language of mathematics."[5]

Heisenberg's pessimism regarding our inability to speak about symmetries except in mathematical terms is not shared by the present writer, for this is what we are about here. Nevertheless, the idealist view propounded by Heisenberg contrasts starkly with the modern materialist idea that all material existence comes from a subatomic particle yet to be discovered. This phantom particle has been called the "Higgs boson" and, according to Paul Davies, was dreamed up in the 1960s as a solution to the problem of how subatomic particles like the electron acquire mass. The Higgs boson was given the title of "the God particle" by American Nobel Prize winner Leon Lederman.[6] This is an interesting title because the explanation for the existence of the Higgs boson refers to the context of a "scalar field" from which the God particle is supposed to be generated. One would think that the scalar field itself should have been given the title of God, as it appears, in this theory, as the fundamental level of reality. However, materialists appear to be stuck on the idea that mass is the irreducible phenomenon of the universe and are therefore inclined to stick with interpretative strategies that ignore framing contexts.

Now while there are instrumental differences between materialist and idealist views on the question of the irreducible phenomena of the universe, both positions nevertheless converge when looking for the "one," the unitary principle that governs the course of the world. But the "one," as an atom or a Higgs boson (materialist) or an Idea (idealist), does not result in a break from dualism or the exclusive ordering by a rational/empirical paradigm. If there is a unitary principle that governs the course of the world, that principle is the *inclusive* principle of symmetry, which is neither materialist nor idealist, but a synthesis of both.

The inclusive structure of symmetry automatically connects matter with ideas because both matter and ideas are made of the same "stuff," which is relationships. This is a nondualistic approach, a holistic view where nothing is excluded from the universal formative and semantic energy of symmetry. Heisenberg's idealist and mathematical understanding comes from the rhetoric of modern rationalism (where God is often seen as a mathematician), which is often opposed to the rhetoric of a modern empiricism used by materialists. The language of both is unable to accommodate the synthesis that symmetrical relationships bring to idealist and materialist positions.

The exclusionist bias of traditional scientific methods, practices, and views constructs a barrier between material existence and consciousness, both between material existence and individual subjectivity and material existence and the divine in cosmic form. These paradigmatic exclusions represents a failure to recognize both the omnipresent nature of consciousness and the full formative and semantic power and divinity of relationships. Given their long period of training in rational/empirical methods, scientists, when faced with the divine, are unable to recognize it.

Nevertheless, since the beginning of this century the concept of symmetry has become increasingly important in the physical sciences. It is a central principle of "new physics," chaos theory, quantum mechanics, and the theories of relativity. Symmetry is crucial to the explanation of the big bang theory of the creation of the universe. The concept of symmetry was significant for Einstein's relativity theories, his space-time transformations, and what mathematicians call a "symmetry group."[7] Symmetry was so central to Einstein's theories that on the occasion of the commemoration of his one hundredth birthday American scientists held a seminar entitled "Symmetries in Science."[8] Symmetry is important in the mathematics of quantum mechanics, a mathematics that describes the specific symmetries of quantum wave-particles like electrons or photons. These wave-particles possess symmetries in the manner that a crystal has symmetry. As Heinz Pagels says, "Quantum particles are like microscopic crystals and can be described completely in terms of their symmetry properties. Symmetry is the key that opens to human understanding the door to the microscopic world."[9] But it is not only the microscopic world to which symmetry is important, for symmetries are all around us. Some scientists have gone so far as to suggest that any system can have symmetry and anything can be a system. For instance, Joe Rosen has argued that

> a system can be a geometric figure, like our square, or a physical entity such as an elementary particle, an atom, a molecule, a crystal, a plant, an animal, the earth, the solar system, our galaxy, or the whole universe. . . . The system might be abstract: the laws of physics, an idea or concept, a mathematical relation, a feeling. In fact, I cannot find a system to which the concept of symmetry is inapplicable.[10]

But while science considers symmetry to be significant for cosmology, physics, and mathematics, the semantic and omnipresent character of symmetry has not been generally recognized. Rather, it is more common in science to reduce and narrow the meaning of symmetry to that of an "interaction" or a single-dimensional "relationship." We find evidence of the traditional scientific narrowing in the manner in which writers of science refer to symmetries. For example, Peat says that the abstract symmetries of elementary particles are quite different from the static symmetries exhibited, for instance, in a rose window, a snowflake, or a starfish. According to Peat the differences are due to static symmetries existing in individual objects, while abstract symmetries exist in the group relationships of particles and in the way mathematics can transform one into another.

This dualist view assumes that there is such a thing as a static object possessing static symmetrical properties that are separate from the observing processes of an individual's subjectivity. It also assumes the common Platonic view that mathematics is a set of ideal forms existing beyond consciousness. From an inclusive, holistic perspective, both these assumptions are seriously flawed. From a holistic perspective there is no such thing as a commonly assumed static object, such as a flower or a stone that is separate from consciousness. Rather, there are only dynamic processes of observation, within subjectivity, where objects like flowers and stones are mimetically constructed and reconstructed from certain structural effects of the environment. As for mathematics, this is simply a set of discourses that can never be separated from consciousness and the communicative processes of mathematicians. Therefore I would suggest that all symmetries, whether abstract or (so-called) static, are actually dynamic and energetic and that in addition they are relationships possessing the power of meaning both for a cosmic consciousness and for individuals.

Peat, who is more likely to lean toward the present view than most scientists, manages to toy with these holistic possibilities. He suggests two possibilities: that the symmetries of elementary particles could be nothing more than descriptive devices and mathematical artifacts that are projected onto quantum matter; alternatively, he asks, "do these symmetries point to a constitutive and formative potential?"[11] He leaves the door open to his own question by suggesting that the symmetries of elementary particles could give a clue to a formative potential. While there may be some scientists who agree with Peat about the formative potential of abstract symmetries, most would probably accept that this potential represent the Platonic and formalist category of mathematical artifacts.

Another popular writer of science who flirts with the idea of the formative potential of symmetries is Pagels. He tells us that "the universe is a message written in code, a cosmic code."[12] Pagels does not decode the cosmic code but implies that it has something to do with symmetry and its consequences like invariance and the law of conservation of energy. Pagels says that the idea of invariance, which was the ancient idea of physical law, "has in modern physics become the idea of symmetry, and the task of modern theoretical physics is to uncover the symmetries of the world. Most of the history of modern physics is the discovery of new symmetries."[13]

I suggest that this view that science looks for new symmetries is misleading. Modern physics may begin by looking for new symmetries, but once glimpsed they are quickly covered over by the codification and formalization of mathematical equations in the manner of some new law of nature. Mathematization is a popular way of "putting a handle on" some difficult or ambiguous piece of nature. This was the approach even taken by the open-minded physicist Heisenberg, who thought that "the central problem of theoretical physics nowadays is the mathematical formulations of the natural law underlying the behavior of the elementary particles."[14] The process of manufacturing differential and scientific apparel for the stuff of divine consciousness may be beneficial for the development of physics, but it creates the illusion that science knows something about the underlying symmetries. This is not so. There are limitations to the knowledge of physics, and these limitations are due to the veil that symbolization creates. In science the veil of symbolization is largely produced by the self-contained domain of mathematics.

Some years ago Sir Arthur Eddington, who was one of the first physicists to fully grasp relativity theory and who became one of its leading exponents, described the reality of physics as "a skeleton scheme of symbols [that] proclaims its own hollowness."[15] This skeleton scheme is made up of what Eddington called "pointer readings": "The physical atom is, like everything else in physics, a schedule of pointer readings. In science we study the linkage of pointer readings with pointer readings. The terms link together in endless cycle with the same inscrutable nature running through the whole."[16]

This skeleton scheme of physics is as much about the internal consistency of mathematics (the rhetoric of mathematics) as it is about the representations mathematics makes of the physical world. Representation and rhetoric are the twin features of all symbolization, and these features are neither innocent or unbiased. The rhetoric of mathematics is no different from the rhetoric of political speeches, for all rhetoric works in the same manner of creating internal consistency through unqualified closures. Unlike Eddington I suggest that symbols do not "proclaim their own hollowness." Rather one of the rhetorical traits of all symbolization is to draw attention to itself while eclipsing the nature of the phenomena it claims to represent. Like the "personalities" on television who eclipse the people they interview, all symbolization promotes its own beauty and style. The mathematization of the physical world is therefore a process of scientific occlusion; a method of diverting our gaze from difficult ambiguities, like the constitutive and formative potential of symmetries, and focusing instead on the skeleton p's and q's of a quantum formula. One of the result of this widespread, inherent occlusion by mathematics has been that science recasts the phenomenon of symmetries narrowly as (1) formalist ideals (Ideas), (2) material interactions (of an inanimate world), or (3) mathematical artifacts.

The occlusion by mathematization of the divine and spiritual potential of symmetries is enhanced by focusing on some of the generalized mechanical effects of a symmetry system. Such is the case with invariance—the proposition that something always remains the same. With this proposition and its related mathematical formula we are no longer troubled about the constitutive, formative, and semantic nature of symmetries. For example, invariance equations like those associated with the statement that "action is always equal to reaction" do not prepare us for discussions about the formative and semantic potential of symmetries—quite the reverse.

The invariance of the "action reaction" relationship is at the center of the science of mechanics and thus seems far removed from questions of formative causation or consciousness. Yet invariance, in general, is an important aspect of the formative and structural potentials of symmetries and the order in which material existence is realized and organized. From a holistic perspective, invariance can be taken out of its narrow, mechanical framework and be recast as an active yet stable process of cosmic life. As such, invariance is just another term for habit: the habits of cosmic organization. We know that the whole complex organization of the universe is intricately managed through stable and complex, recursive orders. From a holistic perspective these orders are not mechanical but are the stable, recursive habits of a cosmic consciousness.

Thus if inorganic matter possesses consciousness generated by symmetry relationships, then the patterns of invariances associated with inorganic matter become the stable behavior, or the habits of material existence. In this manner both physics and chemistry take on the animate features of life. For example, from a holistic perspective, drug and gene research is not so much about discovering the mechanics of different chemical reactions as about analyzing the various habits of behavior of different chemical organizations. This difference leads to a much richer reading of the traditional mechanics of chemical reactions, one that also involves the structures of meaning generated by the organism's formative field.

An organism's formative field refers to the hypothesis of formative causation that Rupert Sheldrake has advanced. Formative causation assumes that morphogenetic fields are physically real. Yet as Sheldrake says, "fields are nonmaterial regions of influence."[17] In biology a developmental or morphogenetic field represents a region of influence that both organizes normal development and guides the processes of regulation and regeneration after the organism has been damaged. Sheldrake interprets these fields of biological influence as having a structure that is determined by the behavior and the forms of "previous similar organisms." The structure of the field thus depends on what happened before. For example, "the foxglove species are shaped by influences from previously existing foxgloves."[18] These fields, which Sheldrake calls "morphic fields," therefore represent the animate collective memory of the species.

Sheldrake proposes that this collective memory of species works on the basis of a resonance; a "morphic resonance takes place on the basis of similarity. The more similar an organism is to previous organisms, the greater their influence on it by morphic resonance."[19] From the present perspective the only step Sheldrake has not taken is to refer to the central role of relations. When saying that the basis of morphic resonance is similarity, in terms of relations, Sheldrake is saying that morphic fields are structured by symmetrical relations. The more symmetrical they are (the less nonsymmetrical), the greater the resonance.

Resonance, as discussed in chapter 1, is seen here as only one of the features of relations. In contrast Sheldrake locates resonance as the fulcrum term of his theory of formative causation. If we were to recast Sheldrake's work by making a conceptual shift from resonance to relations, this slight change would make manifest the strong implications of cosmic consciousness within his theory of formative causation. If this happened, perhaps his colleagues would be even more offended by this further non-Newtonian development.

If we refer to morphic fields in terms of relations, we will also answer the question of a species' collective memory. For example, "How is it possible for an inanimate field to possess a memory"? This is a difficult question unless we know that the relations that structure the field are always relations of meaning. Relations of meaning are relations of consciousness and are thus animate. Morphic fields are therefore animate fields of consciousness that determine the form and behavior of their constituent organisms. That the form and behavior of most organisms is relatively invariant and habitual does not mean that meaning is absent or the field inanimate. Consciousness is habitually habitual as well as creative.

Morphic fields are therefore physically real in the sense that they are nonmaterial fields of relations that are prior to the physical constituents of the field.

Morphic fields thus give meaning, purpose, and direction—as well as organize the behavior and physical shape of the organisms that emerge from these fields.

Sheldrake's morphic fields and Bohm's quantum potential speak of two ways in which cosmic consciousness orders our universe. Bohm's quantum potential refers to the symmetry potentials that organize inorganic matter at a quantum level, while Sheldrake's morphic fields refer to the species fields that organize life at the macro level of organisms. Both fields are structured by symmetrical relations. Both fields have the semantic evidence of a consciousness; of a memory and a "knowing." Both fields are formative in that they are able to develop emergent formations. Both fields transform their own symmetrical relations to create their constituents; in so doing, they pass on some of the symmetry potential of the field to these constituents.

These are some of the features of an animate universe that are erased, ignored, and devalued by the reasoned materialism of a Newtonian and Cartesian perspective.

A Science of Symmetry?

The exclusive methodologies of science divert our attention away from the energetic and complex semantic characteristics of symmetries and in the process deny to us a significant field of investigation. That field is the science of symmetry. This science, if established, would of necessity bring together and integrate a range of presently disparate disciplines, not least among them religion and physics.

If such a science can be developed, if it can be sustained beyond the immediate rejection of the idea by both scientists and theologians, one of the first jobs would be to explore some of the main effects and characteristics of symmetry. This would need to be done so that we are not blinded by "the unreasonable effectiveness of mathematics" in accounting for symmetries. For example, it would be necessary to conceptualize and articulate, in a descriptive manner, some of the characteristics of symmetry before mathematizing them; mathematization should be a secondary function to description. And description would need to be related to the six purposes of symmetry referred to previously: descriptive (or symbolic), constitutive, immanent, formative, structural, and semantic.

These are large and general questions that need fuller exposition than we have room for in this study. But perhaps the first difficulty with the relation of symmetry is the problem of how to conceive of it. Traditionally symmetry has been described in mechanical and exclusive terms, for example, "When a relation holds between A and B, it also holds between B and A." A concrete definition of symmetry is represented by the mathematical equals sign (=) or by the common phrase *the same.* The term *spouse* is an example of a symmetrical relationship, for it implies that the relationship is reciprocal between the two partners.

Symmetries are found in geometric forms, particularly in the repetitive patterns of Middle Eastern mosaics, in the geometric shape of snowflakes, crystals, and starfish and a myriad of other life forms. Symmetries occur in music in the form of the resonance and harmonics. There is also a sound that can be said to be the sound of symmetry. This is the sound of *om,* the ancient Hindu sound of the universe. "The syllable OM, which is the imperishable Brahman, is the universe."[20] Symmetry and its continual repetition (recursive symmetry) is the central feature of the mathematics of chaos theory, in which the output of one mathematical function (iteration) is used as input of the next in a feedback loop that is repeated endlessly. Symmetry is the relationship in all expressions of love, identification, bonding, faith, and religious conviction. Symmetry is the feeling of the heart, that is, the heart as metaphor for love, empathy, and compassion.

Symmetry is therefore both intellectual and emotional; analytical yet spiritual, of the head and of the heart, connecting mundane work with the transcendent consciousness of a God. It carries the common connotation of unity, unification, connection, integration, totality, and linkage and is the major relationship in synchronicity, similarities, and all resonance like periodicity, vibrations, oscillations, rhythms, and cycles.

One of the key features of symmetry is its recurring circularity. Take, for example, the circular linguistic effects inherent in propositions about symmetry. Any propositions about invariance will be formulated in the key terms *same* and *equal,* which are essential to the statement and to their associated mathematical equations. As each of these terms is another way for representing the meaning of symmetry, the concept of symmetry recurs throughout such propositions and formulas. Thus if circularity, or what mathematicians have called recursive symmetry, is an essential feature of symmetries in general, then all energetic systems will be characterized by the resonance of recursive symmetry.

These are common and narrow applications of symmetry. But the divine and formative potential of cosmic consciousness should also be included in our understanding and conceptualization of symmetry. Jung tells us that when "Yahweh created the world from his *prima materia,* the 'Void,' he could not help breathing his own mystery into the Creation which is himself in every part, as every reasonable theologist has long been convinced."[21] This view proposes that all the universe is God and God is in all the universe in every part, for all times: the organization of this universe is therefore God's organization, and the structure of this universe is God's structure. To fully understand the conditions of symmetry is thus to understand and conceptualize God. While this may be something of a tall order, there are many aspects and features of symmetry that are not beyond recognition, understanding, and articulation.

For example, symmetry is an important concept in relation to space and time. Albert Einstein showed the interconnection and relativity of space and time. This should come as no surprise to the student of symmetry who recognizes that both space and time are integrated with each other within the relative conditions of an individual subjectivity that has mimetically reconstructed their characteristics from a synthesis of meaning taken from subjectivity and cosmic consciousness. From a holistic viewpoint, both space and time are nonmechanical; that is, they have two integrated and interrelated features: a background aspect of symmetry and a differential aspect of nonsymmetry. Spatial symmetry is firstly a background void, the *prima materia* upon which the relatively fixed secondary dimensional markers of distance, depth, and width are located. These markers or "pointer readings" should not be identified as the primary, background void of symmetry. Rather, these cardinal markers have emerged out of the void and provide it with some basic differential and explicit meanings. These differential meanings can be added to, for example, by the mathematics of gravitational constants. But no matter how complicated we make these "pointer readings" by mathematization, such differential symbols, along with the cardinal markers of distance, depth, and width, can never become the primary dimension of the background void of symmetry. The pattern of identification that places these secondary features as primary is the false identification of a reasoned materialism.

In like manner the background symmetry of time is the undifferentiated duration upon which a variety of markers can and do produce a variety of differential arrows of time. Music is a good nonmechanical example of the two integrated aspects of the dimension of time. Temporal spaces between the notes represent a background symmetry of duration. This symmetry creates a great deal of the meaning and feeling of all music. The rest of the meaning of music is provided by the differential notes that are established through the complex of symmetrical relationships between notes. These are not two separate sets of meanings, however; they are integrated in that the differential relationship between notes is always established against the symmetrical background of duration.

It comes as no surprise that Einstein should have shown the interconnection of space and time rather than their division, for it is our contention that the emergent characters of both space and time are constructed within individual subjectivity, as well as being located within the cosmic consciousness of the universe. These two "layers" of consciousness are linked by the "normal" symmetry transformations of perception (discussed in chapter 4). The level of structural integration within subjectivity and between individual and cosmic consciousness is high because both "layers" are structured by symmetry and its emergent relations. Space and time are therefore two integrated features within the integrated field of consciousness.

Nor is it surprising that Einstein should have discovered space to be curved. As I have said, one of the key features of symmetry is its recursive circularity. Mechanics operates on what seems to be a flat plane because of the limitation of its spatial and temporal coordinates. Space and time, however, are energetic systems of relationships involving the recursive integration of individual to cosmic dimensions and as such will show recursive characteristics, that is, aspects of curvature and circularity.

Yet we still tend to think of space in mechanical terms, even when we use non-Euclidean geometry. To do so makes no sense at all, for the curvature of mechanical space is like the inside of a ball or some other geometric figure. But we can ask, "What is beyond the ball?" The mechanical answer would have to be another ball-like space! Perhaps a series of spherical or hyperbolic spaces! These are all geometric images within *x* differential dimensions of space that derive from a mathematical language that assumes a differential world where the observer and the observed are separate. If this assumption is denied and the observer and observed are both aspects of a greater whole, that is, are within the recursive and integrated field of meaning and consciousness, then mechanics breaks down and objective geometric figures are irrelevant.

From a holistic perspective the curvature of space involves the recursive circularities of the complex levels of relationships that embed us as individuals into the recursive contexts of the universe. The symmetrical relationships that create this bonding exist physically, chemically, biologically, and culturally, and each of these supports and reinforces the others in complex, circular feedback processes. From this perspective the curvature of space is seen as the manner in which we as individuals relate to the universe, which in terms of meaning and consciousness is cosmic consciousness or God.

From a theological point of view we can say that without God there would be no universe and so no space or time. From the logic of symmetry we can also say that without the symmetries of duration and the void there would also be no space or time. Both space and time have the two integrated aspects of a background context of symmetrical relations from which emerges a set of constituent, differential features (pointer readings). It is our contention that this synthesis of context and constituent is a universal feature and basic to all knowledge and communication. Much of this book will be taken up with the nature of the synthesis of background context with constituent surface feature. Like material existence, space and time are fundamental creation processes, but creation processes that operates with invariance and relative stability. For space and time to occur, the markers of distance, depth, width, and time have to be relatively invariant. If these markers were randomly structured, that is, occurred randomly, then space and time, as we know them, would disappear.

It should be noted that the symmetries that structure the nonlocal, nonspatial interconnections of the void and duration should not be dismissed as nothingness or as emptiness. The symmetries of the void represent an eternal and infinite fullness, a fullness that has created the myriad objects of the universe that appear to us in conscious awareness. This is a fullness of the potentials of relationships of meaning, a fullness of the potentials of cosmic consciousness or God. Even Einstein's law of gravitation is a statement about the fullness of the potentials of the void, for it refers to so-called empty space having certain quantities, called potentials, that obey lengthy equations. These potentials are never named beyond the circular rhetoric that mathematization creates, yet even the cold, abstract notion of an empty space full of potentials is one that resonates with the fullness of cosmic consciousness.

The fullness of the void Jung called the *pleroma,* a term he took from Gnostic theology. It refers to the "abode of God and the totality of the Divine powers of emanations."[22] The mystery of the pleroma Jung described as an eternal and infinite fullness possessing no qualities since it has all qualities: "In the pleroma there is nothing and everything. . . . Therein both thinking and being cease, since the eternal and infinite possess no qualities."[23] The present view contrasts with Jung's only in that we seek to ascribe some qualities to the fullness of the pleroma; these are the general qualities of meaning and relationship and the specific quality of symmetry (and the relations that emerge from symmetry, to be discussed shortly). In other respects we would agree with Jung that in the pleroma there is nothing and everything. The mystery of Jung's pleroma therefore reduces, for us, to the mystery of eternal and infinite symmetries.

Theology is largely concerned with the nature, values, and meaning of our relationship to God. Jung, who was deeply affected by theological and mystical beliefs, spent years writing and studying the nature of the fullness of cosmic consciousness. His works on the psyche, the collective unconscious, the archetypes, the self, and the relationship between religion and psychology represent a series of attempts to explain the formative and divine nature of this fullness. I do not intend to revisit his writings here in any detail except insofar as they provide some examples for the formative energy of symmetry. It would be possible to reinterpret most of Jung's writing in terms of the symmetries they presuppose, but this would be a book on its own. Rather, I want to finish this section on symmetry with some examples from his work that contrast with some of the basic assumptions of mechanics. The reason for this focus is to develop a discourse of integration that draws the reader away form the easy materialist conclusion that relationships are simply physical links and connections rather than active agents of a pervasive and formative consciousness.

One of the assumptions I have been using about symmetry is its ubiquitous universality: it operates at every level of reality at all times. Yet symmetries, while implying the concept of "the same," are always different, in the contradictory manner that all differences are always the same. Since every context will display different symmetrical effects and since every part of the pleroma is a context, we have a multiplicity of symmetrical effects. Some of these symmetrical effects become the empirical world we know through our senses. If isolated from its context of cosmic consciousness, the empirical world appears to be mechanically organized and to obey mechanical laws such as the conservation of energy and the law of invariance. But when recontextualized into the recursive frame of cosmic consciousness, the symmetries of the physical world are no longer just stable, mechanical processes; they take on attributes of the divine.

The stability of cause and effect, for example, should not be taken only as a formal, mechanical "law of nature." It can be viewed in that fashion, but even if we do take on this terminology, we are immediately committing ourselves to the implications of a higher formative order. A "law of nature" always already assumes an order and organization of nature that is imposed from beyond the context of empiricism and the mechanics of cause and effect. It is therefore impossible to speak of the laws of nature without also implying a formative context in which the mechanics of empiricism operate.

The difficulty of empirical neatness does not get any better if we simply do not speak of the laws of nature, but only of principles. All general principles or invariant functions have a context in which they become so, for it is linguistically impossible to state any generalized principle without also implying the context to which it refers. Contexts can, however, be implicit or explicit. An implicit context is not *no* context. It is simply a context yet to be made explicit. The implicit context of all mechanistic laws and principles is the fullness of the symmetries that structure the pleroma. This is the context of cosmic consciousness or God. Thus all symmetrical effects, even empirical and mechanical ones, are features of a divine consciousness.

Yet because each symmetry effect is different, some will indicate the presence of divine consciousness more clearly than others. Jung's notion of synchronicity is an interesting case in point. Synchronicity is an example of a set of symmetrical connections or processes involving an external event and a subjective state that defies mechanical explanation. These types of symmetrical effects are thus very different from the symmetrical effects of mechanical laws. Jung tells us that "the problem of synchronicity had puzzled me for a long time. . . . I was investigating the phenomena of the collective unconscious and kept on coming across connections that simply could not be explained as chance groupings or 'runs.' What I found were 'coincidences' which were connected so meaningfully that their 'chance' concurrence would represent a degree of improbability that would have to be expressed by an astronomical figure."[24]

The classic example of synchronicity told by Jung concerned a patient who was going through a crisis in therapy. The patient was a woman who had a highly rational approach to life that made any form of treatment difficult. On one occasion she had a dream in which a golden scarab beetle appeared. Jung knew the symbolic significance of the beetle: a sign for rebirth. As the woman was speaking about her dream, Jung heard a tapping at the window behind him, and as he opened the window, a gold-green scarab flew in. The patient was so deeply affected by this synchronicity that her Cartesian armor was pierced, and her treatment thereafter improved.

A whole range of synchonistic events have no mechanical explanation. Stanislav Grof, a psychiatrist who has worked with thousands of patients, states that "time and time again I have seen highly improbable accumulations of mishaps and accidents in the lives of people. . . . the only cause we can find for these synchronicities is the capricious interplay between our inner worlds and the physical world outside us."[25]

Physics is normally concerned with causal connections, but symmetries, as the formative agents of consciousness, do not always produce, through the interaction of physical bodies, the mechanics of constant connections through effect, that is, causal connections. A synchronicity is an acausal connection or a meaningful coincidence. A synchronicity involves strong symmetrical connections between subjective and environmental events that are endowed with significant meaning. A synchronicity has little place within the causal conventions of physics. A mechanical view of meaning is usually of a subjective value different and separate from the objective physical world where causal connections operate. Yet meaning was the very value suggested as a feature of synchronicity by the physicist Wolfgang Pauli to his friend Carl Jung in their discussions.

Pauli, who received the Nobel Prize in Physics in 1945, was preoccupied for most of his life with the symmetries in nature and the psyche. He suggested to Jung that meaning was the feature that could integrate the mechanics of causality with the metaphysics of the acausal. This proposal also synthesizes the objective and the subjective, as both rely on the active element of meaning. The nature of a synchronicity comes from the significant meaning attached to a particular pattern of events experienced by an individual. This significance of the meaning is so intense that the divine is seen to have intervened in the process. The connection of events, subjective and physical, is said to be acausal because there is supposedly no empirical connection between a subjective and an objective event.

Yet a connection is made, a synchronic connection structured by symmetrical patterns. From a mechanistic point of view the connection between a physical and subjective event is called a "coincident" or a random parallelism. As relationships are usually seen as nothing but formal properties of physics and carry no meaning of their own, a coincidence is nothing but a statistical calculation. In contrast, from a holistic perspective all relationships are relations of meaning, so all relationships form a continuous field of meaning that includes the cosmic as well as the individually subjective. A synchronicity that happens within a continuous field of meaning is thus a nonlocal connection, an acausal, or more specifically a nonmechanical, connection of symmetry involving the intervention of cosmic consciousness (which must always be present in a continuous field) in the subjective context of the individual. This intervention is seen as such by the person(s) experiencing it, and thus this kind of significance is attributed to the occurrence.

Most of us have encountered strange coincidences that defy ordinary (mechanistic) explanations. I did when I read Heinz Pagels's *The Cosmic Code,* a book about modern science. The last paragraph contains these sentences.

> I often dream about falling. . . . Lately I dreamed I was clutching at the face of a rock but it would not hold. Gravel gave way. I grasped for a shrub, but it pulled loose, and in cold terror I fell into the abyss. Suddenly I realised that my fall was relative; there was no bottom and no end. A feeling of pleasure overcame me. I realised that what I embody, the principle of life, cannot be destroyed. It is written into the cosmic code, the order of the universe. As I continued to fall in the dark void, embraced by the vault of the heavens, I sang to the beauty of the stars and made my peace with the darkness.

On the first page of the book there is a short description about the author, his work and publications. The last sentence says "He died in 1988 in a mountaineering accident in Colorado." Was this a dream of chance, a mere coincidence, or a significant foretelling, a synchronicity? Certainly the symmetry of the dream and the event is a nonmechanical connection of some significance.

Rationalists may charge that acausal or nonmechanical connections admit to supernaturalism, which for many is the same thing as superstition. In answering this charge Eddington has said, "Insofar as supernaturalism is associated with the denial of strict causality, I can only answer that that is what the modern scientific development of the quantum theory brings us to."[26]

Let us conclude this chapter by reminding ourselves how old is this knowledge that the universe is God and that God is in all the universe, in every part, for all times. This holistic view goes back thousands of years before mechanistic science was invented. Hinduism, the oldest religion in the world, tells us continually that "everything is Brahman, that there is nothing but Brahman." These sentiments are echoed today by science itself. "All things are not made of subatomic particles, all things including subatomic particles, are ultimately made of God." This is Ken Wilber writing in the introduction of *Quantum Questions*. Wilber points to the symbolic nature of physics and its radical failure to say anything about the wider context in which it operates. This failure, Wilber says, "paradoxically led so many physicists to a mystical view of the world."[27] For example, physicists like Eddington, Pauli, Schrödinger, De Broglie, Planck, and even Einstein in a limited way were willing to go beyond the symbolic shadows of physics and embrace the metaphysical or mystical.

The only new feature in this old tale is the present proposal that the pleroma, or cosmic consciousness, has some qualities. These are the eternal qualities of vibrational energy, relationship, symmetry, and meaning. These four qualities have a sixfold purpose, constitutive, immanent, formative, structural, semantic, and symbolic. From these qualities and purposes we can say, in summary, that God in the form of symmetries is immanent in everything (there is only Brahman). The semantic energy of this formative vibrational power has meaningfully constructed the material universe from itself, and this undivided universe, which has material and symbolic features, has a coherent and interconnected structure and an organization that obeys the laws of symmetry. One of the laws of symmetry is recursiveness. Thus an individual's relationship to God is recursive. As constituent features of the cosmic context, we are both in God and God is in us.

Chapter 3

Nonsymmetrical Emergence

In this chapter I want to continue to develop the thesis that we can understand individual and cosmic consciousness through the generative relationship of symmetry and its effects. The focus of the last chapter was on some of the qualities and purposes of symmetry. In this chapter the focus will be on the emergent relations of nonsymmetry that produce the realization of material existence. Some comments will also be made in regard to the third set of relations, which are asymmetrical.

An Isomorphic Paradigm

I have argued in chapter 1 that consciousness is constituted by circular processes and should properly be analyzed by circular and holistic methods. The term *holistic* has a multitude of uses, and even though it is defined in chapter 1 as embodying the organizing principles of inclusion and recursive circularity, a further refinement seems necessary in light of the major role relationships play in this analysis of consciousness. A holistic term that seems to emphasize relationships is *isomorphic.*

The term's etymology adequately reflects its use here. The prefix *iso-* is from the Greek *isos,* meaning "equal," while *-morphic* pertains to "shape, or "having a form," and, combining the two, we have, "corresponding to or having the same form." A third influence on the history of *isomorphic,* one that will be stressed later, comes from Morpheus, Ovid's name for the God of dreams, "the son of sleep."[1] In this book the link between sleep and unconsciousness is important: symmetry has an unconscious character. I will be arguing later that the relation of "symmetry" has the characteristic of being unconscious and that an isomorphic paradigm is one that reflects its own unconscious foundations. *Isomorphic* therefore implies an "iso"-"morphic"-"morphia"; *an unconscious* *equality of forms*. An isomorphic paradigm is therefore a perceptual and analytic framework that has at its center the unconscious principle of symmetry.

In the preface to *Chaos Bound,* N. Katherine Hayles uses the term "isomorphic paradigm" to describe and account for the theoretical framework of those physical and mathematical studies that have been called "chaos theory."[2] Hayles tells us that the central value of chaos theory is recursive symmetry and that its contradictory effects are seen in an orderly disorder or a predictable unpredictability. Hayles saw that the notion of recursive symmetry could be applied to both the social sciences and the natural sciences. That insight is appropriated here and made explicit in relation to the structure of consciousness. The concept of an isomorphic paradigm and the recursive conditions that it implies seem to account exceedingly well for meaning and consciousness.

Employing an isomorphic point of view, we return to and look more closely at the relationship of nonsymmetry and its formative role in material existence.

Nonsymmetry

In the introduction of *Quantum Questions,* Ken Wilber tell us that "virtually all physicists believe there are irreducible things (quarks, leptons, gluons) that arise out of broken symmetries."[3] Broken symmetries, in the terms of this study, are nonsymmetrical relationships. Wilber's statement thus indicates that the so-called irreducible material things arise out of nonsymmetrical relationships. In other words, irreducible material reality is reducible to nonmaterial relationships!

Wilber's use of the term "broken symmetries" is unfortunate, as the adjective *broken* indicates that a physical structure has been subjected to a mechanical action that has destroyed, damaged, or modified it. This terminology fits well with a mechanical view of the world, in which symmetries are not called relations and where relations are seen narrowly as passive properties of physical things. For example, if a crystal has symmetry and the crystal is broken, then the property of symmetry is also broken. In this passive, materialistic sense, symmetries are broken by a change in physical structure. This is the point of view of the materialist who holds that the universe can be reduced to irreducible particles like the Higgs boson.

Nonsymmetry is traditionally viewed in a narrow, mechanical, and formalistic manner. We thus speak about nonsymmetry in terms of traditional logic, such that when a relation holds between A and B it does not necessarily hold between B and A. Such a relation is nonsymmetrical and represents a distinction or a difference of some kind. This manner of referring to nonsymmetry implies little connection with the relation of symmetry. Yet nonsymmetry is an integrated relationship in a system of relationships. The prefix *non-* can be used as negation, opposition, absence, failure, and exclusion, but when used with *symmetry* the prefix incorporates all these uses into the more positive concept of difference. In the theories of poststructuralism difference has the special double meaning of deferral and differing, but a discussion of the distinction between the dictionary use of the term difference and Jacques Derrida's term *differance* is not necessary here. From the point of view of this study I intend to accept the general dictionary use of difference in the sense of a "dissimilarity," and more specifically in the terms of the present study, as "nonsymmetry."

I have said that the relation of symmetry is most directly represented by the equals sign (=); the negation of this meaning can be represented by the negation of the equals sign (≠), as in $X \neq Y$. It should be noted that in this representation of nonsymmetry the stroke of negation does not exist by itself as a free-floating negative but reacts to the already functioning meaning of symmetry. It is difficult to imagine how we could represent negation or difference in such a manner as if it acted alone and did not function in response to the already existing background of symmetry. In other words, it is difficult to see how negation or difference can exist without its context. Yet this exclusion of context is the basis of mechanical and formalistic views that narrow the meaning of difference into free-floating negatives. This is the case in the philosophy of language, in which structuralists, poststructuralists, and some postmodernists overvalue difference to the extent that they see it as the sole, yet indefinable, meaning of language.

From an isomorphic view, difference can function only as a relationship that fits within the context of symmetry. This integration is emphasized by the transformation of meaning the term undergoes when as "nonsymmetry," difference is transformed from the free-floating negative into a feature that has emerged from symmetry. This narrow, mechanical discussion on the logic of nonsymmetry can broaden into something much more interesting, for symmetry is the ground of all material existence and the ultimate level of reality. The logical emergence of nonsymmetry therefore fits into the general emergence of material existence from its context of cosmic consciousness.

Wilber's statement that "virtually all physicists believe there are irreducible things (quarks, leptons, gluons) that arise out of broken symmetries" makes more sense if we replace "broken" with *non,* because this term implies an integration with, yet a transformation of, the active agent of symmetry. For physics the term "broken symmetries" operates very much as the term *difference* does for linguistics. Both usages tend to provide a sense of independence, a functional separation. It is our proposal that the so-called broken symmetries of physics are not separate functions, nor are they primary functions. They are like the logical results of the discussion above; they are secondary manifestations of the primary and formative agent of symmetry. They "arise out" of the symmetry of cosmic consciousness.

Clearly there is an ambiguity at work in this discussion of Wilber's statement. Wilber speaks of "irreducible things" arising out of broken symmetries, while I suggest that broken or nonsymmetries arise out of symmetry. Wilber refers to material things, while I refer to relations. Are we talking about two different things? I suggest not. I have suggested that the formative and constituent nature of symmetry can produce the realization of material existence. In other words, material existence is brought about through the advent of nonsymmetrical relations. Nonsymmetrical relations not only mark the distinctiveness of physical objects, they represent the actual agency by which physical objects come into being.

For a physicist's account of the emergence of nonsymmetry from symmetry, we need only turn to the narrative of the big bang. Pagels tell us that

> the primordial fireball was a mixture of all quarks, leptons, and gluons at enormous temperatures corresponding to ultra-high energy. At very high energy the distinction between the weak, electromagnetic, and strong interaction does not exist—they are all unified and have the same strength. The interactions were all symmetrical and equal at the creation, but as the fireball expanded the temperature dropped and the exact symmetry of the interactions became spontaneously broken. With the breaking of the symmetry, the various different interactions became apparent.[4]

For the physicist the element that is significant in descriptions of the big bang is the changing temperature that *causes* the "spontaneous" breaking of the symmetries. Thus the materialist turn of phrase in Pagels's descriptions suggest that perfect symmetry was "spontaneously" broken as the fireball expanded and the temperature dropped. In contrast to this view, consider relations as the active agents rather than the passive effects of material causes. In that case symmetry was not broken spontaneously; rather, nonsymmetrical relations emerged out of the perfect symmetry of the cosmic background of consciousness, and in the process "the various different interactions became apparent." With these changes emerged the different elements of the universe. The difference in terminology relates to the predetermining question of whether the universe is animate or inanimate. "Spontaneously broken symmetries" is the discourse of an inanimate material universe, while "nonsymmetries emerging from symmetries" is the discourse of an animate universe.

I find the materialistic discourses of spontaneously broken symmetries lacking conviction and begging questions like "Why did the fireball come into existence?" and "How did the big bang happen?" The conventional answer to these questions is that the entire universe came from nothingness, period. This is absolutely unacceptable in mechanical and materialistic terms, yet it represent the prevailing wisdom of substance thinking.

The notion of an animate universe makes much more sense even using prevailing descriptions. The sense in Pagel's description is that creation was an emergent process of some significance in which differences emerged from the symmetry of the background context. With these differences, material existence in the form of elementary particles was born. His description is interesting, for it tells us that at the genesis of creation all relationship was symmetrical. This implies that before creation, all relationships were also symmetrical, which is in line with the proposition that cosmic consciousness—God—represents the potentials of symmetry and that these have existed before creation, eternally.

The emergent processes inherent in the big bang are not unique, for they appear to occur today in the continuous transformations involving the creation and destruction of elementary particles.

Pauli's Exclusion Principle

Science now knows that material reality is continually created by the interaction of subatomic particles. These interactions do not arise from individual particles in isolation, but from the "social" environment of particles. This information is well known and I do not wish to add to it. My only contribution is to shift the focus away from the mathematization of particle interactions onto an explicit and active role of relations that always structures the "social" environment of particles and people. Most of science has tended to treat relations badly, as peripheral or as implications not worthy of the full light of analysis.

The interactions of subatomic particles, which I would call relationships, are not random or unorganized but follow a strict, ordered procedure that involves an attraction force and a repulsion force. These two "exchange forces," as physicists call them, are said to be symmetric and antisymmetric, and so in terms of relationships they are symmetric and nonsymmetric. Exchange forces are significant, for they account for the way atoms bind together in particular molecule formations, which led Linus Pauling, among others, to a theory of molecule bonding.

The theory of repulsion (the nonsymmetric result) was developed by Wolfgang Pauli and has become known as the Pauli exclusion principle. The success of the exclusion principle has been said to be greater than that achieved by Einstein's relativity theory,[5] but the problems it solved were technical and of interest to few. The principle of exclusion is not logical or due to causality but is social and acausal and comes from the collective effects of the interactions of particle groups. (Such a collective effect represents a consciousness, an intelligence, a knowing). The Pauli principle explains the periodic table of the chemical elements on a theoretical basis and details the manner in which different chemicals are built up from the behavior and organization of their constituent particles. This is how Peat describes Pauli's exclusion principle:

> Pauli argued that, at the quantum level, all of nature engages in an abstract dance. Moreover all the elementary particles and quanta of energy can be divided into two groups depending upon the type of dance they execute. Electrons, protons, neutrons, and neutrinos, along with other particles, form one group (and engage in an antisymmetric dance) while the other group includes mesons and photons of light (and forms a *symmetric* dance). It turns out that, in the former case, the nature of this abstract movement or dance has the effect of keeping particles with the same energy always apart from each other. However, this *exclusion* of particles from each other's energy space is not the result of any force which operates between them or indeed is an act of causality in the normal sense, rather it arises out of the *antisymmetry of abstract movement* of the particles as a whole.[6]

While Peat lays emphasis here on the "social" nature of these movements, what is unclear in this account (and in many others) is an apparent contradiction in the symmetrical logic that structures the forces of attraction and repulsion and thus the Pauli principle. The logic of this structure is such that when two particles, such as two electrons, have the same energy and try to occupy the same space, they repel each other. This means that two symmetric "forces" repel each other and create what physicists call an "antisymmetric" function. An antisymmetric function is a mathematical way of referring to a nonsymmetrical set of relationships. In other words, two symmetric "forces" have led to a nonsymmetric result, which is called an exclusion principle.

The significance of the formula of "two symmetries equal a nonsymmetry" is that it appears to be an example of a general principle of emergence, not just for the creation of material existence, but also for the realization of conscious awareness. I will have more to say about conscious awareness later, but for now some of that discussion can be foreshadowed. Conscious awareness represents a secondary set of functions within individual subjectivity that have emerged from the primary set of symmetrical and unconscious operations. This emergence of conscious awareness (for example, in attention) derives from the interaction of symmetrical relations, produced by the nervous system, coming into contact and resulting in the creation of a nonsymmetrical set of relationships—in more conventional terminology, differences and discriminations.

As Freud has implied, conscious awareness is noted as a perceptual system of differences, distinction, and discriminations. Each of these terms is understandable as a reference for nonsymmetry. Conscious awareness thus emerges in the nervous system of individuals as the result of systems of nonsymmetries, and these nonsymmetries have in turn been produced from the interaction of two systems of symmetric relationships. These processes will be more fully elaborated in relation to excitation and inhibition in chapter 7.

The general principle that says "the interaction of symmetries will produce nonsymmetries," which we would argue governs the emergence of all material existence, also provides for the emergence of conscious awareness. These two developments, one physical and one psychic, come together when we observe, through the processes of conscious awareness, material substances as physical objects. In other words, the processes of observation mirror the processes of physical emergence. Objects therefore represent, not irreducible substances with an "in-them-selves" quality, as some materialists would have us believe, but the result of twin symmetrical transformations functioning in unison. This double set of iterations, involving the emergent processes that construct conscious awareness and physical entities, structurally mirror each other in their creation but also in their unqualified completeness (or what I would call their closure) that provides the object's sense of separation and exclusion. The symmetry of these dual emergent systems perhaps explains the strength (for some) of materialism as a irreducible reality.

A general principle of emergence in which the interaction of symmetries result in a set of nonsymmetries accounts for the spiritual belief that everything comes from nothing, or as the Hindu scriptures say, "In unity diversity arises."[7] It is my proposition here that the creation of material existence emerges out of the background context of an "active intelligence" of the pleroma; from the nothing of the symmetry of cosmic consciousness. The fullness of the pleroma is thus the fullness of the potentials of everything, as the theory of the big bang implies. Jung said that "in the pleroma there is nothing and everything."

David Bohm

A further development on the theme of the emergence of material existence comes from David Bohm's concept of the implicate and explicate orders.

Bohm developed his "causal interpretation" of quantum mechanics by first accepting Niels Bohr's orthodox version, which has always been called the "Copenhagen interpretation." Bohm became unsatisfied with Bohr's formalist view, which tended to erase the reality of the quantum world and replace it with "only an abstract mechanical description" of that world. In reaction, Bohm developed his "causal interpretation," which proposed a real quantum universe describable in terms of the quantum potential that extends into the generalized notion of the universe as a interdependent, holistic web of instantaneous interrelations.

This undivided universe is made up of two movements, or wave functions. He called the first movement the "implicate" order and the second the "explicate" order. The implicate order has an enfolding movement and the explicate order an unfolding movement. Bohm explains that the term *implicit* is based on the verb *to implicate,* which derives from the Latin meaning "to fold inwards" (as multiplication means folding many times).[8] The implicate order is seen by Bohm as a universal embedding order that enfolds the information and structures of the world within itself.

In describing this undivided universe Bohm refers to a cosmic determinism that is very close to the present concept of a cosmic consciousness. "We see quite generally that the whole universe not only determines and organizes its sub-wholes, but also gives form to what has until now been called the elementary particles out of which everything is supposed to be constituted."[9] Here Bohm is referring to a constant universal process of creation and destruction that is organized and determined in specific ways. For example, particles are created from an unfolding brought about by the explicate wave function and destroyed by enfolding produced by the implicate wave function.

The emergence of material existence thus derives from the explicate order and its destruction through the implicate order. Both orders, however, are not simply abstract, mathematical functions but refer to actual movements. These movements are capable of ordering, organizing, and determining the subsystems that make up material existence. These movements are composed of a series of interactions, in other words, a complex of relations. Within these capacities, which are the capacities of relations, we see the movements and organization of a cosmic consciousness or intelligence.

The terms "implicate" and "explicate" orders are both seen here as synonymous with implicit and explicit meaning. This being the case, we can say that cosmic consciousness and individual subjectivity are structured in similar ways by implicate or implicit meaning as well as by explicate or explicit meaning. These two orders thus indicate that consciousness is a two-system structure: a primary system of implicit meaning and a secondary explicit system that has emerged from the first. This two-system structure aligns with the general psychological view that subjectivity comprises unconsciousness as well as conscious awareness. Thus a further psychological elaboration on this two-system structure, which develops from an isomorphic perspective, is that the implicate represents unconscious, implicit meaning and the explicate represents the explicit of conscious awareness.

Material existence thus emerges in conscious awareness through observation in the same manner as it has emerged as physical forms within cosmic consciousness, that is, through an explicate wave function. There is therefore a double wave function involved in the emergence of all material existence: one at the level of cosmic consciousness and one at the level of subjectivity. Both explicate wave functions have the same structure, and both have emerged from the ground of the implicate order, the implicate order of subjectivity therefore representing a subsystem of the implicate order of cosmic consciousness. If we accept that perfect symmetry produces perfect unconsciousness then the degree of fidelity achieved in this double implicate/explicate movement is high enough so that its division is absolutely hidden from the normal processes of observation when observing.

Bohm's implicate and explicate orders come together with our earlier discussion of the Pauli exclusion principle through the structure of symmetry and nonsymmetry. Not only do the implicate and explicate orders equate with meaning, with explicit and implicit meaning, they are also structurally synonymous with the movements of symmetrical and nonsymmetrical relations. This connection implies that the implicate order is a hidden symmetrical movement, or wave function, from which emerge all particles that are nonsymmetrical forms of the explicate order or wave function.

It needs to be noted that the explicate, nonsymmetrical forms (particles) are nonsymmetrical in relation to the symmetry of the implicate ground from which they have emerged. This nonsymmetry of all particles is an ontological nonsymmetry distinct from the nonsymmetric function of some particles. The emergence of these explicate, nonsymmetrical forms therefore represents the effects of a wave function prior to the operations of the Pauli exclusion principle. For example, as the Pauli exclusion principle operates at the level of particle interaction, these interactions represent interactions of an explicate wave. At this level, particles participate in symmetric and nonsymmetric (antisymmetric) dances. These movements occur within the confines of an explicate wave function and are entirely in keeping with the symmetric and nonsymmetric potentials of the prior wave function from which they have all emerged.

Some particles, such as mesons and photons of light, will continue to participate in symmetric functions, while others, such as electrons and protons, will function nonsymmetrically. Yet both these movements are the natural extensions of the symmetric and nonsymmetric potentials from the originating implicate order. What Bohm would call the determinism of the wave function is carried through all explicate levels, so that this potential becomes a superquantum potential that affects all explicate levels in the entire universe. Structurally, there is a recursive symmetry at work in this superquantum potential that extends, by a series of wave functions that enfold and unfold, through every level. This is a description of what Bohm has called the "wave function of the universe."

The wave function of the universe is not like an ocean wave that finally breaks and dissipates on the shore. The wave function of the universe is a cosmic function (not mathematical) and it is holistic and circular (not extended to a finite point). Bohm has said that "broadly one could say that through the human being, the universe is making a mirror to observe itself."[10] This is a constituent/context, a circular and holistic relationship, a process in which the wave function of the universe, once began, has the potential to affect every aspect of the universe. But such a wave function is not a mere physical function but has all the determinism and formative potential of cosmic consciousness, which is then reflected back to itself in the subjectivity of individuals.

Apparently many people find Bohm's notion of the implicate order difficult to understand. John Hogan, in a tribute to David Bohm's life and work, quoted the mathematician Sheldon Goldstein as saying, "I've never met anyone who really understands the implicate order."[11] This inability to come to terms with the world of symmetry potentials, of the hidden, the implicate, the unconscious of the background context, is a common reaction among linguists and philosophers and, apparently, mathematicians. The traditions of the explicit have been so strong that many who are trained in its imperatives find it difficult to imagine the nonmanifest as the nonmanifest, the implicit as the implicit, or the unconscious as always the background context, preferring instead to see the implicit as a kind of difficult or ambiguous explicitness.

We find this attitude emphasized in John Searle's *The Rediscovery of the Mind,* which rails against such notions as the unconscious and the background and the possibility that the nonmanifest may not be able to be made fully explicit. The exclusiveness of the explicit is also emphasized in all formalist linguistics. Many may find the concept of the implicate order difficult to accept as an essential aspect of knowledge and the world, but from the present isomorphic point of view, the implicate order is helpful in understanding the principles of symmetry and emergence. We should, therefore, not be dissuaded by this traditional bias in favor of the rationally explicate from pursuing an understanding and appreciation of the character and potentials of the implicate, infinite consciousness.

In order to describe the relationship between the implicate and explicate orders Bohm used several physical examples. He employed the analogy of the hologram because of its propensity for each of its parts, in some sense, to contain the whole. A hologram is a three-dimensional image constructed by splitting a laser beam in two, directing one half onto a photographic plate and bouncing the other half off an object and then onto the same plate. The two halves of the beam meet at the plate and the light waves of each half interfere with the other. When a laser light is directed through the processed plate, the interference wave patterns captured by the plate reveal a three-dimensional object. The holistic significance of a hologram becomes apparent, however, if we cut a piece from the plate and shine a light through the fragment. The image produced will not be a section of the original object but will be an image of the whole object. The meaning of the whole object is thereby contained within (in Bohm's terms "enfolded" within) each section of the photographic plate.

The hologram is an excellent example of a holistic process, but it can also be seen as an example of an isomorphic model. As an example of a more general model, holograms have some explanatory power. However, if used as a primary model of explanation, as for example the neuroscientist Karl Pribram has done, then holograms, I believe, are useful in limited but interesting ways. Like other complex systems, holograms are themselves amenable to isomorphic interpretations that emphasize the role of relations. For this reason they help us come to terms with the way in which meaning is iterated within an isomorphic system.

Recurring symmetry is a transformation of symmetry, or what mathematicians call a "symmetry transformation." In the hologram, recursive symmetry occurs spatially in every part of the image where the whole is contained within every part. Bohm's unfolding of information would thus seem to operate by way of symmetry transformations, which is another way of speaking about the iteration of meaning but also about the emergence of nonsymmetry from symmetry. The hologram is a complex example of emergent principles, from the symmetry of parts-to-whole to the emergence of the nonsymmetries of the parts and, in addition, we would speculate, from the symmetry of the coherent light of the laser to the emergence of the nonsymmetry of the interference that arises from the interaction of the two laser beams. What emerges from these intricate transformations is explicitness in the form of nonsymmetrical relations involving space, time, and light.

Bohm also used the example of a television broadcast as a process involving the implicate and explicate orders. In simplified form the transmitted radio waves carry an enfolded or implicate order of the visual images being broadcast. The function of the television receiver, at the other end, is to explicate this order, that is, to unfold it in the form of new visual images that replicate the ones already broadcast. The enfolding and unfolding of meaning is therefore a fundamental process in television broadcasts.

These analogies of a hologram and television are somewhat limited as examples of the universal, implicate order operating at a subatomic level. To express the dynamic and all-pervasive nature of this quantum order Bohm coined the term *holomovement.* The holomovement is the nonmanifest wave function of the universe out of which all manifest and explicit entities arise. Capra saw Bohm's approach as studying the order of the holomovement, "not by dealing with the structure of objects, but rather with the structure of movement, thus taking into account both the unity and the dynamic nature of the universe."[12] The holomovement is the universal implicate order of the cosmos, in agreement with Jung's view of the pleroma as the "totality of the Divine powers of emanations" in which there is nothing and everything, and is consonant with the Hindu belief that the "world and cosmic consciousness are synonymous."[13]

Bohm's stress on the holomovement and the implicate order turns the tradition of the explicit on its head. By regarding them as having priority over the explicate order, Bohm is reversing the normal values awarded to the explicit by the traditions of reason and empiricism. Instead of giving priority to the formalism of reason as well as the empirical belief in the explicitly material, for Bohm and the present writer, these traditions represent a secondary set of processes that are constructed out of the primary background field of implicate symmetries.

The traditional rational, empirical, and mechanistic worldview is of separate objects moving in space and time, that are measurable by symbolic means. Bohm remarks that "generally speaking, the laws of physics have thus far referred mainly to the explicate order. Indeed, it may be said that the principal function of Cartesian coordinates is just to give a clear and precise description of [the] explicate order."[14] This is a good summation of rational and empirical approaches that assume science can only work with explicit, differentially mechanical, or reductive features that are best measured by Cartesian coordinates or some other exclusive means.

Inherent in Bohm's holomovement is thus a holistic and isomorphic view of the world that contrasts with traditional views. For example, Bohm is careful to point out that the implicate order should not be conceived of as an expression of the explicate order. That is, it "is not to be understood solely in terms of a regular arrangement of objects (e.g., in rows) or as a regular arrangement of events (e.g., in a series). Rather, a total order is contained, in some implicit sense, in each region of space and time."[15]

The general features of this description assume an isomorphic paradigm. Like the implicate order, an isomorphic perspective is not a new form of circular reason. The holistic, recursive "logic" of an isomorphic framework is not the linear logic of a rational cause and effect. Implicitness cannot be readjusted into a new form of explicitness, for the implicit will always remain as a foundation background context for the explicitness of any constituent. And finally, the orientation of reason is one that values exclusive differences, while an isomorphic perception privileges the recursive integration of differences. Isomorphism cannot become a new form of reason.

Bohm has also suggested that these two orders of the implicate and the explicate should not be viewed as a binary pair but as a the result of a series of dialectical interactions in which the implicate, which is fundamental, constructs the explicate, which is a secondary manifestation of the implicate.[16] By way of example and to emphasize this point, Bohm refers to the dialectical way in which music is comprehended. When music is played at its proper speed, we do not hold one note in awareness and then another and then another in sequence; rather a number of notes are present in awareness together in various "degrees of enfoldment,"[17] that is, in various degrees of unconsciousness and conscious awareness. It is the simultaneous and varied degrees of awareness of all of these notes that constitutes the sense of the unbroken flow of the music.

The nonrational term *asynordinate* can be used to refer to the integration of Bohm's implicate and explicate orders, which incorporates the integration of symmetry, nonsymmetry, and asymmetry. Such integration also includes the interrelationship of cosmic consciousness and "world-appearances." *Asynordinate* is taken from Bohm's work and refers to the double nature of the explicate order involving all explicit units, entities, and identities. This double nature involves the integration of *(a)* a foundation context of hidden, implicit meaning, with *(b)* the explicit, visible, constituent features.

An asynordinate relationship is analogous to the relationship between the waves and the ocean. The waves are explicit forms that emerge out of the implicit ocean, but they are not separate from the ocean, just distinct. Waves are asynordinate because they have explicit forms that are structured by the implicit context of the ocean. In the same way symbols are asynordinate, for they are explicit forms that arise out of a hidden context of implicit meaning. Physical entities and processes are asynordinate because they are realized, first, through their emergence from cosmic consciousness and, second, through the hidden symmetry transformations of perception. Conscious awareness is also asynordinate in its content as well as its structure, for it always has a foundation of unconscious subjectivity and has arisen out of that hidden foundation. *Asynordinate* represents both the dialectical results of emergence and the structural relationship that the hidden and implicit always has to the explicit and differential.

An asynordinate relationship is usually not equally balanced between the implicit and the explicit but is weighted in favor of the implicit, or in Bohm's terms, in favor of its "degrees of enfoldment." This is the case with discourse, subjectivity, and, I would suggest, the universe. We see this bias toward the implicate order in the manner in which Bohm describes the relationship between the implicate and the explicate orders. His asynordinate view of this relationship is also reflected in Freud's approach to the relationship between the unconscious and the conscious. In the *Interpretation of Dreams* Freud remarked that the "unconscious must be assumed to be the general basis of psychic life. The unconscious is the larger sphere, which includes within it the smaller sphere of the conscious. Everything conscious has an unconscious preliminary stage."[18] Not surprisingly, Bohm's view of consciousness was similar: "the easily accessible explicit content of consciousness is included within a much greater implicit (or implicate) background."[19]

While there are many similarities between Bohm's view of the implicate and explicate orders and the approach taken here, there are also differences. These differences relate to the subject matter a physicist confronts and the present attention given to consciousness. Bohm was concerned about the order of things at a subatomic level and therefore focused on the concept of order. In *Science, Order, and Creativity* Bohm and Peat draw attention to the breakdown in communication between Niels Bohr and Albert Einstein in the middle of this century. This break, which is reflected today in the distance between relativity and quantum theory, was represented by Bohm and Peat as one based on different notions of order. In particular, "Bohr believed that the order of movement of a particle would admit ambiguity while Einstein felt that such a possibility was too absurd to contemplate."[20] Bohm and Peat then go on to relativize the concept of order by talking about the characteristics of new orders of science. The implicate and explicate orders are therefore orders that relate directly to the physical world "out there," and only indirectly to the processes of observation and consciousness.

When Bohm comes to address consciousness and language directly, he sees them as just another system like matter to which the implicate order relates. Bohm's view of matter was to see it as distinct from consciousness, but he did believe that in general they are "different aspects of the one overall order."[21] But these different aspects are subordinated to the general notion of order, which transcends matter and consciousness. This decontextualization of order as a formal process represents a traditionally rational view that relegates consciousness to a secondary position. In contrast, I see consciousness as the primary phenomenon that primarily constructs order and disorder, reason and isomorphic connections, chaos, randomness, and predictability. In this isomorphic sense the concept of order is subordinated to consciousness, not the other way around, as Bohm proposes. The effect of this isomorphic view is to take consciousness not as a derivative of some larger impersonal order of the universe, but rather as the central organizing and energetic formative system incorporating all orders. It is therefore not so much a question of how consciousness and matter are connected to some higher order. They are connected automatically and holistically within cosmic and individual subjectivity through the emergent and formative relationships of symmetry, nonsymmetry, and asymmetry (asymmetry is discussed below).

Gordon Globus has seen a related problem in Bohm's implicate order. Globus's difficulty arises out of the way he sees the holomovement relating to individual subjectivity. Globus argues that if the holomovement has ontological primacy over the world by virtue of it generating the world, then the power of unfolding the implicate order stays with the holomovement. The power of unfolding cannot be given to individual subjectivity, which is a derivative of the holomovement.[22] This argument points out quite neatly the basic difficulty of beginning with isolated concepts like the physical world and working backward to other isolated concepts like observation and consciousness. This approach appears to be Bohm's along with science generally. The problem that Globus suggests, however, does not arise if consciousness has a isomorphic and holographic structure, one in which cosmic consciousness inheres in individual subjectivity. This is not an either/or situation but an integrated, asynordinate both-ness, together. Unfolding the implicate order is therefore both a cosmic and individual preoccupation.

In concert with the present approach, Bohm made no distinction between language and consciousness and saw meaning as enfolded in the structure of language as well as in thoughts and feelings. He suggested that thoughts enfold together and that a train of thought is actually a process of enfoldment of a series of implications.[23] But Bohm does not articulate how the implicate and explicate orders operate in language, nor does he detail how these two processes function in consciousness. To have done so would have meant, first, developing a discourse analysis that went beyond the reductionist principles of poststructuralism, and, second, transforming Freud's theory of the individual's unconscious into a general theory of linguistic and social unconsciousness. As a physicist, Bohm did not pursue these paths.

The difference between the present isomorphic approach and Bohm's rests on a commitment to isomorphic implications. The present approach views Bohm's implicate and explicate orders, not as transcendental processes, but as orders of consciousness, cosmic and individual, that isomorphically determine the order in which the universe has evolved and continues to develop. Bohm's implicate and explicate orders are therefore principles of consciousness, principles by which material existence emerges from cosmic consciousness and principles by which we observe and give coherence and structure to the world. Such a coherence is only possible because the structure of consciousness is isomorphic with the overall structure of the universe. I agree with Bohm's implicate and explicate orders and his general notion of the holomovement. But they represent, not a sovereign province of the physical world, but emergent structures of individual and cosmic consciousness.

The present approach also goes further than Bohm by proposing that these overall organizing principles of the implicate and the explicate orders, that of "enfolding" and "unfolding," have a structure, and that structure is constructed by the semantic, formative, and emergent relations of symmetry, nonsymmetry, and asymmetry.

To summarize this section we can say that nonsymmetrical relations do not stand alone as explicit and formal negatives but are the emergent results of animate, symmetrical transformations. Nonsymmetrical relations not only mark the distinctiveness of physical objects in conscious awareness but represent the actual agency by which physical objects come into existence as substances and then are observed. This view may be drawn from descriptions of the big bang, from the interactions of particles, and from Bohm's implicate and explicate orders. To see it, we must be willing to see relations, not simply as the incidental effects of material causes, but as active and formative agents of a cosmic consciousness.

Asymmetry

A final word should be said about the third and final tier of emergence in this story of relations. The third tier is the development of asymmetrical relations. Asymmetrical relations are formed from a network of nonsymmetrical relationships embedded within a series of symmetrical relations. Such a network is a system of some kind. In quantum physics a system would be called a wave function. A system is always composed of an integrating, background context in addition to various visible and distinct, constituent, surface features. Like the waves on the ocean, the asymmetrical forms of any system (which is also always a system in conscious awareness) are given context, structure, and meaning by the implicate and symmetrical order of cosmic consciousness.

Asymmetrical relations are conventionally spoken of in terms of irreversible relationships, such that if a set of relations holds between A and B they never hold between B and A. The set of relations represented by such terms as *before, after, greater, more, less, above, part, whole,* and many more are relations of asymmetry.[24] However, asymmetrical relations are never singular relationships because they are always formed from the "social" interaction of the other two relationships. In addition, asymmetrical relations cannot stand alone and apart from the other two relations of symmetry and nonsymmetry as asymmetry emerges from these.

All systems are asymmetrical but also asynordinate; that is, all systems are composed of a unconscious hidden background and an explicit foreground of distinct features. All systems have the potential for transformation, either by way of symmetry transformation, for example by unfolding the hidden meaning contained within the system itself, or by transforming the system into a constituent of a larger system. All systems represent the "social" aspect in any set of interactions (relationships) involving particles or people. Any system or wave function will exist, not independently, but within the context of some larger system, and this process goes on infinitely until all systems interact with the infinite of cosmic consciousness.

The smallest unit of conscious awareness is always a system, as is the largest unit of conceptualization, which is God. A single concept, percept, image, or symbol is therefore not an ontological or material entity but always a system of relationships. Terms like *difference* and *similarity* represent simple expressions of systems. Yet in common to all systems, both these terms refer to connecting contexts that are wider systems that provide much of the meaning for these terms.

If every system is always connected to other systems, which are then connected to others, and so on, then this infinite extension often does not appear as such. This is because the facility of focus in conscious awareness reduces and limits the extension of systems. Such reductions or closures, which have grown out of the necessary mechanical limitations of conscious awareness, come from a pragmatism that ignores or erases connecting systems. Such a pragmatism is the basis of a reasoned materialism that not only limits the interconnection of systems but also limits meaning, for much meaning resides in the wider contextual systems.

A system represents the third emergent level of relationships—the other two being symmetry and nonsymmetry. Systems occur at the level of conscious awareness; therefore, it is impossible to hold in awareness a single and isolated nonsymmetrical relationship. Differences always come in groups, and groups are systems. Pauli's exclusion principle is therefore concerned with systems (of interactions) that exclude other systems and is not concerned with isolated single exclusions of nonsymmetry. The terms *function* and *wave* are often used in physics, and these can be interpreted as referring to systems of relationships.

Even though materialism is pervasive in technological cultures, it is difficult, if not impossible, to isolate material reality from that which is different from the nonmateriality of systematic relations. Scientists do not possess a supernatural ability to engage directly with the physical world and so would do well to try this exercise. All investigations of the physical world are subject not only to the limitations of our technological measuring instruments but also to the limitations of our conscious awareness, which operates only on the basis of systems of relationships. This is the limitation of a world perceivable only in terms of a tertiary level of emergence, it is the limitation of life and the framing restrictions of a physical world mediated by interconnected systems. This means that observation is always organized in terms of systems, and such a subjective bias is unavoidable even when studying quantum mechanics.

Because conscious awareness functions on the basis of systems, all knowledge is about the interaction of systems. This is another way of saying that all knowledge is concerned with the relations of symmetry, nonsymmetry, and asymmetry. From an isomorphic perspective we should strive to perceive the world as systems and move away from the traditional and exclusive focus on single or binary constituents. This is the traditional Aristotelian rational and empirical view that prescribes the ontological primacy of matter and substances and denies the unconscious, the hidden, and Bohm's implicate order. In contrast, an isomorphic view of relationships provides a role for the implicit, binding unconscious and emergent potentials of symmetry.

Summary

The general concern of this chapter has been the emergence of nonsymmetrical relationships and the distinct role they play in an undivided universe. The features of the universe that traditionally appear most disconnected from consciousness are material objects. I have argued here that material objects are created and realized through nonsymmetrical relations that emerge from the symmetry potentials of cosmic consciousness. Cosmic consciousness, or what Bohm calls the holomovement, realizes material existence through unfolding, or what has been referred to here as "the principle of emergence." This is the principle that states that nonsymmetry (distinctions) emerge from the interaction of symmetrical relations. Material existence is realized (by unfolding) as distinctions within the context of the symmetry of the holomovement (cosmic consciousness).

Material existence therefore represents an emergent expression of the symmetry of cosmic consciousness. The relationship individual subjectivity has to material existence will be the subject of the next chapter.

Chapter 4

Substance and Subjectivity

In this last chapter of the section I want to explore some of the isomorphic implications of the relationship between physical substance and individual subjectivity. These two come together at the point we call the environment. Some concept of the environment is important to any nonmaterialistic theory of consciousness. The environment cannot be ignored. To do so would imply that madness and insanity are irrelevant to such a theory. On the other hand, the environment cannot become the determining cause of consciousness. If this were the case, we could not escape the reductive and materialistic effects of mental states being seen as brain states. Both these extremes are unacceptable to an isomorphic theory of consciousness. How then should we view the environment isomorphically?

Generic Subjectivity

An isomorphic view of subjectivity is that it represents a local example of cosmic consciousness, having an asynordinate structure that generates meaning and energy and possesses some limited formative abilities. The structure of a particular subjectivity will be treated the same as another for two reasons: (1) roughly the same functions are performed by every nervous system and (2) the relational structure of each subjectivity will be identical. This structure is generated by the relations of symmetry, nonsymmetry, and asymmetry. Generalizations about subjectivity are therefore possible across cultures as long as they are concerned with the universals of relationships and do not take up the promotion of specific cultural values. As I am using English here to describe the universals of consciousness, it is not possible entirely to exclude some cultural values.

A key feature of an isomorphic approach is the assumption that every perception and all knowledge is filtered by the subjective and that there is no division between subjectivity and the physical world. This is a bounded yet unlimited view of subjectivity that is quite different from the traditional Western view that locates consciousness usually as the featureless entity of the "mind" or "cognition" separated from the physical world. This traditional view rests on the schizoid belief in "objectivity," which assumes that we can stand outside the processes of consciousness while viewing the world.

One of the main features of an isomorphic approach is the acceptance that every "approach" rests on tacit, underlying assumptions. One of the underlying assumptions of the "objective" rational and empirical approach is disconnection, for example, the disconnection between observer and observed, mathematics and subjectivity, or language and thought. In contrast, a key assumption of an isomorphic approach is that everything is interconnected and integrated. For example, subjectivity is a field of meaning that constructs everything we know, feel, sense, and experience in any way. The physical world is therefore not a separate dimension divorced from subjectivity but an integrated aspect of that subjectivity. This means that everything is subjective, everything is a system of meaning, and everything comes within the boundaries of our subjectivity.

The implications of this isomorphic view are that the observer and the observed are part of the same subjective field of meaning. It means that science is conducted, and always has been conducted, within the confines of experiences of the subjective. It means that mathematics, like language, is a subjective, discursive experience and not some kind of Platonic essential form. This is a perspective that agrees with the ancient Greek motto Know Thyself, for it is only through a knowledge of meaning and the structure of our own subjectivity that we can know the world at large.

While an isomorphic view of consciousness holds that everything is subjective, not all subjectivity is simply opinion or related to freedom of choice. To begin with, there is a great variety of unconscious bodily sensations, systems, and operations that work automatically and over which we have little conscious control. But in addition to these prereflective systems there are also conceptual constructs through which symbolic representation occurs and through which we do have some choice. The field of subjectivity is therefore not simply a three-dimensional space of explicit awareness—which is the conventional Cartesian view of the cogito. Rather, all subjectivity has an asynordinate architecture involving involuntary and largely unconscious bodily sensations, desires, perception, or practices that underpin a limited freedom of choice concerning the selection and combinations of symbols.

Another feature of the isomorphic approach is the denial of the either/or of dualism and its replacement with a "both-ness." For example, a both-ness is evident in the integration of mind and body, of emotion and intellect, and of observer and observed. In the case of the environment a both-ness implies both subjectivity and matter together in a composite unity. How is this possible?

The Environment as a System

There are at least four ways to view the environment:

1. An objectivist position simply holds that the "observer" exists independently from the environment and that it is possible to innocently experience unmediated, empirical "reality." This view accompanies the usual mechanical and materialistic assumption made in classical science.
2. In opposition to this is a solipsistic position that holds the reverse to be the case; that empirical reality is totally constructed by the mind. This idealist position is associated with the notion of "mental matter" and the British philosopher Bishop Berkeley.
3. Yet another perspective has been advanced by the biologists Humberto Maturana and Francisco Varela, who have argued that the environment acts only as a "trigger" affecting the "autopoietic," that is, the self-creating self-organization of the nervous system.[1]
4. A final position is the isomorphic one, some of the features of which are implicit in earlier discussions.

The simple, exclusive, and dualism of the first two options is dismissed here, while the third choice of the environment as a nondeterminate "trigger" is discussed in terms of its limitations for a theory of consciousness.

Because of the present emphasis on relations, the question arises of what to call the environment so as not to slip into objectivism and ignore subjectivity, or slide into the solipsistic camp and erase environmental interactions. We need to devise a set of mediating terms for the environment that balances both these positions. In line with the present isomorphic approach I use the phrase system *type* to describe the environment, which is similar yet distinct from the *system in general* of subjectivity.

As Rosen has pointed out (see chap. 2), anything can be a system. If this is the case, then, from an isomorphic perspective, any system will also always be a synthesis of different types and the system in general of subjectivity. In this sense the environment, its patterns of substance and its objects, represents distinct and different system *types,* while the whole system of subjectivity represents the *system in general.*

The concept of system type refers to a pattern of differential relations that emanate from the cosmos but that are recreated, by transformation, by the perceptual processes of the nervous system. System types can originate as social forms, such as different types of codes in discourses, or they originate from physical conditions that we are then able to label by such names as *chair, tractor, building, solar system,* and so on. It is the differential patterns of a system type that will always produce different systems, but the structurality of each of these systems—the system in general—will be produced by the three isomorphic relations that structure subjectivity. This is to say that every system as such derives its structurality, its principles of structure, from the isomorphic structure of consciousness, which in turn is reproduced by the human nervous system as subjectivity. In contrast, specific types of system derive their character from social interaction, as in the case of the symbols of discourse, or from the myriad, explicated forms of the cosmos.

From an isomorphic perspective, we can never know the physical "in itself," nor does it have meaning for us "in itself." We can only know the physical in terms of type of system within our system in general of subjectivity. In this study, therefore, when I refer to environmental objects, the identity, implied by the term *object,* represents a composite synthesis, a system type within the system in general of subjectivity.

In the sense that isomorphic relations provide the structurality of every system, the structure of the system in general is provided by the semantic relations of symmetry, nonsymmetry, and asymmetry. This implies that every system, observed, interpreted, or imagined, whatever the type, will always be contained within subjectivity and be structured by those three relations. And again, in the terminology of mathematics, any system will always have the potential for symmetry transformations, for a transformation represents a relation of nonsymmetry enclosed within the identity of the system.

This isomorphic approach therefore disagrees with Locke's suggestion that the mind is like an empty slate, a tabula rasa. In contrast, the system in general of subjectivity will always be a system full of symmetrical relationships that are relations full of the potentials for symmetry transformation—which is another term for learning. This is even the case with newly born infants, whose subjectivity does not yet involve the symbols of talking but whose system in general is nevertheless a plenum and fullness of mostly unconscious meanings and potentials.

Symmetry of Perception

The coupling of system *type* with the *system in general* of subjectivity is an organizational structure necessary for the organism's survival responses to its environment. Yet the unity in this coupling provides a set of boundary conditions in any act of perception. This boundary inhibits *(a)* distinguishing between the several possible system types during any single act of perception; and *(b)* distinguishing between the two "layers" of consciousness, that is, between subjectivity and the substances of cosmic consciousness.

The first inhibition is evident in the use of what psychologists call an "ambiguous figure," one drawing that can generate two images, but not at the same time. The second inhibition is referred to by Varela, who asserts that "all states of the nervous system are internal states, and the nervous system cannot make a distinction, in its process of transformation, between its internally and externally generated changes."[2] If it were otherwise and we could distinguish between these two, that is, between what the eye sees as a percept and the unmediated object, then we would have an entirely different kind of nervous system and be a very different species.

This unity of the two layers of consciousness (in perception) has traditionally led to the dualist views of either "objectivism" or solipsism. Some philosophers, such as Heidegger and Merleau-Ponty, have disagreed with this dualism, arguing that the separation of subject and object denies the fundamental unity of being-in-the-world, that the interpreted and the interpreter do not exist independently, and that we participate in the shaping of our world and the course of our joint history. The biologists Maturana and Varela ostensibly agree with this middle position. Varela remarks that "the successor to objectivism is not subjectivism, by way of negation, but rather the full appreciation of participation, which is a move beyond either of them."[3] Varela and Merleau-Ponty call this dialectical middle position "participation," while Heidegger describes it as "being-in-the-world." Each of these writers denies the simple, traditional binary choice between objectivism and subjectivism and chooses instead some kind of position in the middle.

Both Maturana and Varela see their approach to perception as a salient corrective to the "common sense" of objectivism. The objectivist position sees perception as concerned with information that is assumed to exist independently "out there" and that can be partially selected by an "in here" process. When external information is ignored by the "in here" and instead a series of internally generated inferences are projected "out there," this projection is usually considered by objectivists to be an illusion or, in serious cases, hallucination. This objectivist distinction of "reality" versus illusion provides us again with the common binary options. Both Maturana and Varela have challenged this neat approach by proposing that we must study perception from an autopoietic position, that is, from within the closed organization of the nervous system itself and not from the traditional position of a mapping of objective "reality" that tries to compare what is outside with what is going on inside.

In order to understand their position on perception it is necessary to remind ourselves of some of the features of autopoiesis. An autopoietic system is a biological concept and represents a "self-perpetuating," "self-sustaining," and "self-creating" system characterized by an organizational closure of its constituent processes.[4] Maturana describes an autopoietic system in the following terms:

> A dynamic system that is defined as a composite unity as a network of productions of components that, a) through their interactions recursively regenerate the network of productions that produce them, and b) realize this network as a unity by constituting and specifying its boundaries in the space in which they exist.[5]

In broad terms this description seems to refer to many of the processes of an isomorphic subjectivity, for there are many similarities of terms and descriptions between an autopoietic and an isomorphic approach to the nervous system. For example, biologists who are concerned with autopoietic systems look at "structural transformations under organizational invariance."[6] In isomorphic terms this can translate to "the symmetry transformations that occur within the *system in general* of consciousness."

While there are similarities between the present isomorphic view and an autopoietic approach, there are also significant differences. These differences center around Maturana's and Varela's rational view of the nervous system. (Both Maturana and Varela have specifically written about the nervous system as a closed network.)[7] In addition, perhaps because they are biologists, they have said little about meaning, discourse, or consciousness. In any study concerned with the relationship between cognition and neurophysiology these are serious omissions. I suggest that because of these omissions both writers have automatically fallen back on the traditional rational paradigm to describe what is essentially a dynamic, nonlinear, recursive set of processes. For example, the view that "man is a rational animal"[8] hinders the recursive logic of autopoiesis. It does this subtly in their argument in favor of a third alternative that rests on a disagreement with an objectivist position and a specific denial of solipsism. While claiming the middle ground of "participation" and stressing the closure of the nervous system, the rational framework they employ causes them to continually slide into solipsistic conclusions. This slippage, I suggest, is due to the lack of any paradigmatic alternative to fill the space between objectivism and subjectivism.

The term "participation" is not theoretically complex enough to provide that framework. In any case, both writers adhere to the sight lines of reason that continually throws up binary alternatives and exclusive distinctions, thereby working against a middle, integrated position. The rationalism that frames their view also leads them to postulate the legitimacy of the biologist "observer." From an isomorphic position argued consciously, there can be no observer as such, only a participator in the circular processes of interpretation. An "observer" never escapes the processes of his/her own nervous system, nor the structure of meaning. The so-called observer is therefore always a second-order observation, representing an observation of an observation.

Thus the strength of the reason that informs their common focus on the autopoietic system forces Maturana in particular to accept certain closed and solipsistic conclusions even while he denies such solipsism. For instance, Maturana states that

> communicative and linguistic interactions are intrinsically not informative; organism A does not and cannot determine the conduct of organism B because due to the nature of the autopoietic organisation itself every change that the organism undergoes is necessarily and unavoidably *determined by its own organisation.*[9]

In this comment there is no place for environmental influences, for "every change" that the organism undergoes is "determined by its own organisation." This solipsistic conclusion arises from the closed logic of a rational approach. Thus if nervous systems are organizationally closed, then from an *observer's* point of view they are separate and different from their environment and therefore instructive and determining interactions cannot take place. Maturana argues that "the distinction between internal and external causes in the origin of the changes of state of the nervous system can only be made by an observer that beholds the organism (the nervous system) as a unity."[10] But here we have a problem with the general opposition of a "closed system" and the notion of the observer. This difficulty is confusing and provokes a tendency to slide into solipsism, in the sense that "reality" is constructed solely by the organizational closure of subjectivity.

Given the notion of the "observer," we have, therefore, established the "real" environment outside the nervous system. Yet this environment is strongly denied by the implications of the organizational closure of the nervous system. Here we have two perspective paradigms, a rational in conflict with an isomorphic. The rational paradigm has been explicitly used by both Maturana and Varela, while the isomorphic paradigm is implicit in their general concept of autopoiesis. Thus the conflict for both Maturana and Varela is that nervous systems cannot be both beyond closure (as implied in the concept of "observer") and at the same time organizationally closed.

Their emphasis on the organizational closure of the nervous system, in the absence of a strong theoretical middle perspective, leads inevitably to the devaluing of environmental effects that operate outside the system. I take a representative sentence from Varela's *Principles of Biological Autonomy* to illustrate this point: "the nervous system operates as a closed network with no inputs or outputs, . . . its cognitive operation reflects only its organisation. . . . information is imposed on the environment and not picked up from it."[11] In this comment we have a solipsistic tendency to close off effects from the environment while emphasizing internal organization and construction. Even if we accept Varela's point that "inputs" and "outputs" are simplistic and problematic terms that imply the simple transference of "objective" properties from "out there" to "in here" and therefore should be denied, his statement nevertheless closes off environmental effects to the system in general of subjectivity.

In developing their case against the objectivist position, Maturana and Varela describe some examples of sensory perception that they claim cannot be understood as an input process. An input process is one whereby a series of effects in the "outside" environment is mapped "inside" by the brain. To illustrate his disagreement with this simple and naturalistic view, Maturana uses the analogy of a pilot landing a plane in heavy fog using only the mechanisms of instrumental flight. When congratulated on a perfect landing, the pilot says, "I did not fly or land; I only manipulated certain internal relations of the plane in order to obtain a particular sequence of readings in a set of instruments."[12]

This tendency to "throw out the baby with the bath water" is given further support by Varela in relation to "size constancy" in perception. Varela argues that size constancy is not a function of independent features of the object or scene, but rather is a function of a correlation that takes place within the nervous system. The perception of a three-dimensional world is commonly and traditionally analyzed in terms of size, distance, depth, and other components that are believed to be independent of us. Size constancy is analyzed by "objective" means in terms of distance and size of the perceived object, as if these three factors represented a direct and independent set of environmental features. Varela questions these assumptions and analyzes size constancy "as a neural process of invariance that takes place independently of any 'features' of the environment."[13]

Using the "Ponzo illusion" (an illusion of distance created by the apparent convergence to two lines), Varela argues that "we do not see distance."[14] Rather, the effect of seeing distance "corresponds to a process that takes place completely within the nervous system, independently of any features of the environment."[15] His conclusion is, therefore, that an organism does not "extract perceived distance as a characteristic feature of the environment but generates it as a mode of behaviour that is compatible with the environment through a process of invariant compensation of disturbances."[16]

This interpretation downgrades the role of the environment to only "a source of perturbations independent of the definition of the system's organisation, and hence intrinsically non-instructive; they can trigger, but not determine, the course of transformation."[17]

From the present point of view, Maturana's and Varela's theory of autopoiesis does not sufficiently allow for the role, structure, and determinism of the environment that is the cosmic "layer" of consciousness. One difficult effect of this significant devaluation is that their theory is unable to account for the difference between normal perception and hallucination. In terms of subjectivity this denies the whole arena of madness and insanity, which must be part of any theory of consciousness. Such omissions must be considered as a fundamental flaw.

Mimetic Reconstruction

The disagreement I have with this interpretation is that from the point of view of meaning and consciousness the environment is much more than a "trigger" or a "perturbation." It presents a very definite source of system type. These are explicate meanings that are mimetically reconstructed by prereflective processes of subjectivity in the forms and patterns we know as the environment. An isomorphic view of the environment therefore gives it larger determining power than an autopoietic view. Varela's argument comes down on the solipsistic side of the system in general of subjectivity while playing down environmental effects as "triggers." But such an approach lacks explanatory power to describe the significant differences between "normal" perception and hallucination or how we are embedded in the cosmos or even the structure and organization of meaning and consciousness.

From an isomorphic perspective system types can and do determine outcomes. For example, types of words determine concepts, types of subjects studied by students determine an individual's later occupation, types of childhood experiences play a large part in determining the course of personal histories, and in general types of environments determine physical behavior patterns and perhaps conceptual patterns. It has been found in psychological experiments that the lack of variation in types (in the form of sensory deprivation) can induce disorientation and, over long periods, insanity. System type is therefore an important and determining category, unlike the concept of a trigger, which is nondetermining. The analysis here is therefore driven by valuing *both* the system in general as a system of meaning and environmental types that can and do determine outcomes.

The prereflective transformations that occur within the processes of perception when the cosmic and the personal interact represent changes that connect the implicate and the explicate order of both cosmic consciousness and subjectivity. In order for the explicate order of cosmic consciousness to become visible to us it has to be transformed by our subjectivity back into the unconscious implicate order of subjectivity (the unconscious processes of perception), to be then transformed again into the explicate order of conscious awareness (as percepts) to become the visible environment. The "visible environment" is therefore a conflation of two dimensions of emergent processes; the explicate order of cosmic consciousness and the explicate order of subjectivity (see chap. 3).

Between the symmetry of these two explicate orders is the implicate order of our subjectivity that transforms one into the other. The "visible environment" is thus mediated, first, by the implicate conditions of its own emergence, which among other things involves the transformation of the non-spatial into region and locality, and, second, by the implicate order of our subjectivity. The reality of the environment is therefore a fourfold emergence (the enfolding of two asynordinate structures) and is far from innocent and unmediated. It is the reality of a series of transformational processes involving the implicate and explicate orders that can never be observed directly because we are always already part of these processes.

In this participating sense the mimetic unities of the explicit environment are always closed, in that there are no indications of differences of space, time, and structure involved in any act of perception. When we "see" an object like a table, the conflation of the explicit order of subjectivity (in which the percept exists) with the explicate order of cosmic consciousness is not differentiated in any way. Nor could it be, because intervening between these two is the unconsciousness of the implicate order of subjectivity. There is therefore a simultaneity of time, space, and situation, where the "outside" is no different from the "inside" and where the physical type represents the same transformational result as the percept.

From this point of view the environment can be thought of as the necessary, prereflective, projected truths of sense perception.[18]

Simultaneity

The problem of the simultaneity of time, space, and situation in respect to perception raises a number of questions. One such important question is: how similar or different could environmental conditions be from our percept of them? The nervous system and the relations of meaning it produces are isomorphic and not *rational*. This means that any recourse to the "observer" who has God's position from which to observe the environment directly is illegitimate. Questions about the similarity or difference of object and percept are not questions about a relationship with the environment so much as about the limitations of the rational framework that is being used to frame these questions. For example, a rationalist perspective will assume there is an inside and outside of subjectivity and that it is possible to know the environment separately from subjectivity. When this distinction is attempted, the rational result can be a commitment to objectivism, or a slide into solipsism, or a contradictory combination of both.

From an isomorphic perspective the question of correspondence between "in here" and "out there" is not an "objective" question but one about the worth of our approach. As symmetrical relations are fundamental to the isomorphic approach but inimical to the mechanics of reason, we say that not only is there a correspondence between cosmic conditions and our involuntary, perceptual constructions of these conditions, but that this unity, due to its lack of differentiation, is one of almost perfect symmetry. We could, for the sake of description, call the explicate order of cosmic consciousness a *type 1* system, and our involuntary prereflective, symmetrical transformations of that type a *type 2* system, and the closure of their interaction a *type 1/2* environmental system. A *type 1/2* environmental system is a composite unity, a system constructed from the structure of subjectivity as well as from the asynordinate structure of the universe.

This view of the environment does not naturalize it. Any object or entity can, by reflection, be seen as a system. In this endeavor we are often assisted by a range of technologies that can extend the boundaries of our sense perceptions. Such extensions, however, do not change the limits or character of our prereflective perceptual processes. Nor do they change the implicate and explicate orders of subjectivity or of cosmic consciousness. Rather, scientific instruments can extend explicit information and learning by developing ways in which more meaning can be unfolding from the wave function of the universe. This is surely the direction of all scientific endeavor: an explication of the hidden, implicate meaning of the cosmos.

When Varela says that we do not see distance but construct it in our nervous system, then from the present perspective it should not be taken that there are only spatial factors in our heads. From an isomorphic perspective, if there were no such thing as distance in the cosmos, the "distance" constructed subjectively and projected onto the world would hardly "fit" the world and be symmetrical with it. In other words, the "distance" we construct within our nervous system must be symmetrical with spatial conditions already existing in the universe. There must be structural identity because the identity of perceived space with physical space would not take place if it were otherwise. This should not surprise us if we have already accepted that we are always embedded in nature.

As for the question of environmental determinism, within the symmetry of perceptible type there is always a transference of meaning. For example, there has to be a transmission of implicate meaning by the prereflective processes of perception. The category of type is not created within subjectivity but is recreated—not arbitrarily, but in conformity with the implicate meaning that has come from the cosmos. In this sense cosmic influence represents the volumes of hidden meaning that are recombined and selected in conscious awareness as "the environment." The sheer multiplicity of different environmental forms that we "see" in any daily scene is so overwhelming in complexity and design that it is amazing how the implicate order of subjectivity can reconstruct these continuous scenes (which change by the fraction of a second) with such explicitness.

Finally, the basic character of the symmetry inherent in type 1/2 environmental systems constructed in prereflective perception should therefore tell us something about the world as well as something about our subjectivity. It should also point up the similarities in life and subjectivity. For example, it is possible to view the life cycle of an organism as a series of symmetry transformations. The processes of birth, growth, and death of an individual represent a series of unfoldings of an explicit, differential organization and then an enfolding, in death, back into the implicate order of the universe. These transformations, covering approximately seventy years, in the case of humans, are self-similar to the rapid transformations that take place in subjectivity when we read or listen to music. The process of reading involves a series of explicit meanings emerging from the implicate order of unconsciousness that then enfolds back again when conscious awareness moves on. These are the transformations that are occurring in your subjectivity as you read this sentence.

Constructed Realism

From an isomorphic perspective the environment can be represented by the concept of constructed realism. Constructed realism may seem like a literary term that describes certain kinds of fictional discourses, but this is not my use here. The phrase *constructed realism* applies to the prereflective level of subjectivity, not to the symbolic level of representation. It therefore refers to a mimetic presentation that functions automatically and involuntarily in respect of the creation of environmental conditions within the general framework of subjectivity. Constructed realism implies an isomorphic set of processes, the kind of nonrational processes hinted at by Merleau-Ponty and expressed by Langer when she writes that "things have an independence—but not an absolute independence; they are in-themselves for us."[19]

Constructed realism does not imply an absolute freedom to construct the environment as we like, nor, of course, does it argue for the simple materialistic determinism held by "objectivists." Rather, it argues for the reality of "physical" substance that is both constructed and already determined. This is a reality that is not independent or fictional, but a composite unity existing both in subjectivity and cosmic consciousness simultaneously. We can no more construct uniquely and independently the letter *a* than any other object. But then, the letter *a* no more exists independently from its realization in subjectivity than any other object. Thus when a nervous system is operating in a "healthy," nonhallucinating manner, the structure of the letter *a* will always be determined, to an overwhelming extent, by the structure of the system type 1, which impinges on the body from outside it.

For Maturana and Varela the space between objectivism and subjectivism was filled by the term "participation." But the structure of this participation involves a fourfold transformation of two asynordinate systems interacting with each other. The complexities in this prereflective transaction of meaning go far beyond the notion of participation. In contrast, I have called this whole transformative set of processes constructive realism.

Projection

The construction of the environment, of type 1 and type 2 systems, while taking place within the processes of perception, nevertheless often produces a sense of transcendence through the mechanism of projection. This projection is an effect of a simultaneity of situation involving the symmetry of type 1 and type 2 systems. Because this symmetry seems absolute, the differences of space and time are erased, and so the actual location of the perceiving processes themselves can never be reflectively taken into account. There is therefore a nonlocality in perception that manifests itself in the form of projection. We therefore have a sense of "seeing" the world and projecting it "out there," where it is separate from ourselves. Contrary to this sense of "out thereness," we are actually reconstruct the environment within perception. But what we construct has a structural symmetry with what is outside our subjectivity. Therefore, what is projected by symmetry is structurally symmetrical. There is, therefore, a double symmetry at work in "seeing" the world.

The sense of transcendence created by the act of projection should not be dismissed as simply an irrational illusion. It may be an illusion, but from an isomorphic position it represents a projection necessary for the daily, pragmatic survival of interacting with the environment. This pragmatic necessity is given reinforcement in the mechanics of materialism, in which a difference between "out there" and "in here" is always assumed as a given. I suggest that if this possibility of an irrational illusion is seriously contemplated, it will lead to a schizophrenic uncertainty of situation (in here or out there?). In addition, such reasoning will locate the mechanism of projection not as the legitimate category of "necessary illusion," but quite the reverse. From a rational position "projection" is, in principle, a derogatory term.

To some extent the mechanism of "projection" has earned this pejorative status when contingent projections, in the form of reflective discourse, are referred to as necessary, as for example when paranoia is apparent, or delusions of grandeur diagnosed, or even when the naked emperor is seen to have clothes. In each of these cases projection has been universally condemned. Yet in each of these cases the problem has been not projection per se but contingent projections that have, at some degree, been called necessary. The important distinction here is therefore between the necessary projections of prereflective subjectivity and the contingent projections involving representations constructed in reflective consciousness. It is a significant difference in the level of subjectivity, of the nonrepresentational and the representational.

Constructed realism, which is always pragmatically useful at the prereflective level of subjectivity, can become problematic at the reflective and representational levels of subjectivity. Thus, the prereflective projection of an environmental situation provides us with perceptual certainties. These certainties are necessary to perform all our daily tasks and adapt to our environment. A perceptual certainty, for example, means that we do not doubt what we see while seeing it. Without prereflective certainty of situation we would be unable to perform any task. When this certainty is seriously undermined, by either hallucinogenic drugs or the disorientation that may stem from serious repression, sensory deprivation, or perhaps even from long and compulsive training in the imperative of a "superrationalism," we become paranoid and/or we hallucinate. From the standpoint of an isomorphic perspective, hallucinations are signs that representational projections are so severe that they have disrupted and unbalanced the prereflective projections of the perceptual system. Disrupted in the sense that the "normal" certainties that come with constructing our environment are disrupted, to the point that the realism of the environment is put into question. At this point the individual can be said to be "ungrounded" and "living in a fantasy world."

In this condition, constructed realism turns to constructed nonrealism or delusion and hallucination. In this condition also, the individual is experiencing, at the level of conscious awareness, too much alienation and fragmentation. In delusion and hallucination there is therefore an uncertainty of situation, an uncertainty that should "normally" accompany reflective consciousness but that invades the lower, prereflective levels of perception.

The difference between the necessary projections of prereflective subjectivity and the contingent projections of reflective consciousness is therefore a distinction between normal perception and hallucination.

Conclusion

From an isomorphic view the symmetries of perception indicate a series of implicit and unconscious meanings connecting the nervous system with the rest of the universe, with cosmic consciousness. The symmetries of perception imply that the nervous system is structurally and organizationally very much *of* the world and not separate or different from it. It indicates that we both construct the world as well as having these constructions determined to a large degree by the meaning of the cosmos. We are free to construct a world, but not free to construct it as we want. The effect of this isomorphic position is that there is no doubt that we are intimately and ubiquitously affected by our environment because we continually re-create it. This means that as witnesses to violence we becomes a participator in the action, but it also means that as witnesses to beauty we are also part of that beauty.

Part 2

The Context of Subjectivity

Chapter 5

Meaning and Consciousness

"I am interested in meaning because it is the essential feature of consciousness." So stated David Bohm some years ago in an interview.[1] If we agree with Bohm, we must describe and define meaning and how it relates to subjectivity. The purpose of the present chapter is therefore to sufficiently elucidate the meaning of meaning, to make it understood in terms of the second "layer" of consciousness, which is individual subjectivity.

Information

Meaning is not always seen as the essential feature of consciousness. Even Bohm, who has said that "mind and matter are inseparable, in the sense that everything is permeated with meaning,"[2] and that meaning is enfolded within matter, nevertheless backs off attributing consciousness to nature. He suggests, rather, that "everything has a kind of mental side, rather like the magnetic poles."[3] To account for this nonpantheistic, pantheism, Bohm coined the phrase "active information" to refer to the manner in which particle movements are determined. Active information relates to the "mindlike" quality of particles and their reactions to each other, as if they "know" what the others are doing and act accordingly. This knowledge is not conveyed by mechanical forces or dynamic interactions of the conventional kind but represents the "social" consciousness of particles.

Bohm contrasts active information with the more common use of information as a passive attribute of some physical system. The distinction made in chapter 1 between the passive and the active roles of relations parallels Bohm's distinction. Active information is thus a metaphysical feature of the physical world when it displays nonmechanical, mindlike behavior.

If we replace active information with meaning, what is the result? Bohm says that "everything is permeated with meaning." What is the difference between active information and meaning? Clearly not much, for both are used in a metaphysical sense, referring to a metaphysical reality. The traditions of our intellectual training are important, however, for the different emphasis of these two terms. The instrumental use of *information* refers to mindlike behavior. Mindlike behavior is an analogy for mind, but analogies, we know, are never more than that. Information, then, is an analogy for meaning that is an essential feature of consciousness. But such terminological caution is inappropriate for a theory of consciousness.

If we accept that meaning is an essential feature of consciousness, then information refers to several levels of abstraction removed from meaning. In other words, for information to be information presupposes symbols and a reflecting consciousness able to code and decode the information. Information, whether passive or active, is never a primary feature of consciousness but always a secondary effect of a reflective conscious awareness. The instrumental use of information reverses this priority and builds in a homunculus (a little reflecting consciousness) to any proposition about information. In other words, when we use the term information we assume that a reflecting consciousness is an aspect of mindlike behavior that in turn is an analogy for mind!

Bohm was a physicist and not a cognitive scientist, so his instrumental use of the term information can be forgiven. If information and all its derivatives like *computations, analog* and *digital data, coding* and *decoding* were removed from cognitive science and replaced with *meaning* and its derivatives, much of the intellectual superstructure of this most instrumental of sciences would vanish. At present cognitive science seems to represent the science of analogy, for it deals only with the analogies of consciousness, mindlike behavior such as artificial intelligence, sensory coding and decoding, neural networks, and so on. Not all cognitive science is equally distant from consciousness. For example, the cognitive semantics of George Lakoff[4] attempts to shift the focus back to consciousness, but this approach suffers from the underlying difficulty of leaving consciousness and meaning undefined. To have anything relevant to say about consciousness, cognitive science must first confront the central feature of consciousness, which is meaning.

The Meaning of Meaning

In 1923 *The Meaning of Meaning* first appeared. In the course of their wide-ranging review, the authors, C. K. Ogden and I. A. Richards, described sixteen types of meaning. These sixteen essentially came down to two: meaning was (1) an intrinsic property and (2) an unanalyzable relation. The ambiguity of these overlapping definitions and the inability of Ogden and Richards to clarify their own definition foreshadowed a confusion that continues on in linguistics, cognitive psychology, and cognitive science today, so much so that meaning, the central element of these disciplines, has remained largely undefined.

The famous linguist Emile Benveniste had this to say about meaning:

> At this point the problem that haunts all modern linguistics arises: the relationship between form and meaning. Many linguists would like to reduce it to the notion of form alone but somehow they cannot succeed in freeing themselves of the correlative, meaning. What has not been attempted in order to avoid, ignore, or expel meaning? It has been useless; this Medusa's head is always there at the centre of language, fascinating those who contemplate it.[5]

These words are as apposite now as they were Benveniste wrote them in the 1930s. Why is meaning so problematical, so difficult to look at directly—like a Gorgon's head? Linguists have tended to tackle the problem as Perseus did when he approached Medusa, by looking at her reflection in the mirror of signs. Meaning then appears manageable, as a function of the way signs relate to each other, or the way words are used. But the question of what meaning looks like has remained a fascinating mystery.[6]

A more recent approach to meaning is to be found in *Semiotics and Language: An Analytical Dictionary* by A. J. Greimas and J. Courtes, who have this to say: "Although it is a property common to all the different semiotic systems, the concept of meaning is undefinable."[7]

It should be noted that the kind of meaning these writers are referring to is the general concept of the meaning of meaning. It is not meaning as particular interpretation, for example, as in the meaning of something, of a sentence or a sign. This distinction, having a traditional linguistic and semiotic context, is maintained here but expanded into the realm of consciousness and given the generative power of a formative hierarchy involving primary and secondary meaning. Primary meaning refers to the primary conditions or the initial processes of meaning. From an isomorphic view, primary meaning, the meaning of meaning, is the formative, immanent, omniscient, and omnipresent relations of symmetry. Secondary meaning always emerges from the symmetry potentials of primary meaning.

Secondary meaning is always generated by semiotic activity, for example, by signs, through discourse and communication. Secondary meaning always involves reflection and interpretation, for example, the meaning of a sign or the meaning of a particle reaction or the meaning of a text or the meaning of life. The distinction between primary and secondary meaning is important to grasp, for it is the difference between the general symmetrical conditions of the universe, which are primary and which underpin all physical and subjective activity, including communication, and secondary meaning, formed and produced by semiotic and symbolic systems.

Secondary meaning is often referred to by the question "What do you mean?" or the phrase "what is meant by." Primary meaning always involves the context of the presymbolic, the ineffable structure from which symbols, words, and sentences have emerged. These are the differences between the symbolic that has cultural form (secondary meaning) and the innately ineffable, universal, and infinite primary meaning of symmetry. Secondary meaning, which is always conscious meaning, is a double and a recursive system.

This double recursive system is always asynordinate and never a dual or binary set, with primary meaning separated from secondary meaning. An asynordinate system is not a linear hierarchy. The divine and formative power of symmetry does not run on railway tracks; it is transformative and recursive. The recursive, asynordinate structure of meaning implies that secondary systems produce secondary systems of meaning, while primary systems (symmetry potentials) produce secondary systems. The asynordinate structure of meaning also implies an asynordinate and transformative field where primary meaning is always integrated with secondary meaning.

Within individual subjectivity, primary meaning transforms into secondary systems that then interact with other secondary systems to generate interpretative and semiotic systems of secondary meaning. The activity of semiotics is, therefore, brought about by a series of interactions between asynordinate systems, for example, the interaction between *(a)* the use of a language system (English, German, Japanese), *(b)* the cultural context of the expression involving genre, convention, and discipline, *(c)* the personal history of the individual, *(d)* the specifics of time and place. Other distinctions may be discerned in specific interpretive acts, but whatever the set of systems that make up the combination of interactions, these interactions will always be secondary ones.

Once we have said that secondary meaning is always semiotic activity (linguistic and discursive systems), we need to remind ourselves that semiotic activity is always asynordinate, having its foundations firmly fixed in the symmetry of primary meaning. Without this reminder we tend to fall into a reductionist and formalistic view and begin to believe that all meaning is *caused* by semiotic activity. This reductionist view may then be formalized even further into the belief that meaning has a single and explicit dimension and that it arises from symbolic representation and specifically only from logical propositions, other propositions being considered meaningless. Variations on this positivist and reductionist view have pervaded the traditional Western philosophical history of meaning.

Some Conventional Approaches

As Benveniste reminds us, the question of meaning has tended to be ignored, avoided, or "expelled" in linguistic areas where grammar and vocabulary are under analysis. Similar comments can be made about semiotics and the philosophy of language. It is perhaps not surprising, then, that no acceptable linguistic or semiotic definition for meaning has ever been found. Morris, for instance, avoids the term *meaning* because, as he puts it, "accounts of meaning usually throw a handful of putty at the target of sign phenomena."[8] Greimas and Courtes tell us that "the concept of meaning is undefinable" and go on to say that meaning can be approached naively or intuitively and considered in terms of operations like transcoding and internationality, but that to do so will introduce a whole range of (apparently unacceptable) metaphysical presuppositions.[9]

Where Benveniste views meaning as an integrating function of the linguistic unit, Wittgenstein's practical approach to meaning was to see it as a matter of use. Meaning rather than epistemology was central to Wittgenstein's view of language, and in his later writings Wittgenstein identified meaning with use: "The meaning of a word is its use in language."[10] According to Wittgenstein the use of a word depends upon the social and linguistic situation. Meaning and use thereby become almost interchangeable, and sometimes he ran them together as in the remark that "the ostensive definition explains the use—the meaning—of the word."[11] The linguist Michael Halliday appears to accept Wittgenstein's view of meaning as use when he proposes that meaning is function in context,[12] that is, linguistic units function always within some social context as a medium for exchange. From this social perspective, meaning thus derives from the functional (use) of linguistic units within any given cultural context.

To the question "What is the meaning of meaning?" Wittgenstein and Halliday would have no answer other than to restate a description of "use." Colin McGinn suggests that "insofar as Wittgenstein has a positive account of meaning, it is an account the chief purpose of which is to act as an antidote to mistaken or misleading conceptions of meaning."[13] In other words, Wittgenstein had only a functional description of meaning. Other writers have been more disposed toward a positive view: C. S. Peirce, Charles Morris, and Max

Black have referred to anything that produces meaning as a sign, while yet another theorist has used such difficult terms as "bearers of meaning" to indicate signs that carry meaning.[14] While various notions of the "sign" can differ in their design and application, the overall drawback in talking about signs in relation to meaning is that signs will only help us understand *how* traces of meaning are woven within a system of signs, not *what* comprises the traces in the first place.[15]

Structuralists following Saussure take meaning to be the relationship between the signifier and the signified, for example, the images and concepts produced by words. This is not the case for poststructuralists. Derrida appears to accept Saussure's view that "in language there are only differences"[16] but in deconstructive writing we are presented with the difficulty of defining "difference" as *differance* (a neologism for deferral and difference).[17] For Derrida, meaning circulates around *differance,* which links absence with presence, but the meaning of *differance* is indefinable, for it "is no longer simply a concept, but the possibility of conceptualisation." And again, "Differance is the nonfull, nonsimple 'origin'; it is the structured and differing origin of differences."[18] These statements indicate among other things that for the deconstructionist difference/differance (what is accepted here as the relation of nonsymmetry) is *the* essential ingredient of meaning.

Derrida accepts the indefinable and ineffable mystery called *differance* as the qualified "origin" of meaning and text without worrying unduly about the nature of its ineffable character or, on a wider scale, the necessity for integrating linguistic units together, a function that difference is unable to perform. Derrida accepts that difference itself represents an ineffable dimension of language when he remarks that *differance* eludes vision and hearing, that it occupies "an order that no longer refers to sensibilities."[19] According to Derrida this order resists the opposition between the sensible and the intelligible, an order beyond truth "in the occult zone of a nonknowing."[20] If these passing descriptions add up to the inexpressibility of *differance,* they do so on the basis that *differance* is a kind of mysterious "play of signifying references which constitute language."[21] Such statements tell us little about meaning and *differance* other than to indicate the difficulty of describing them from a formalist and reductionist position.

A more traditional linguistic approach is taken to the question of meaning by Keith Allan. In his *Linguistic Meaning* Allan identifies seven historical answers to this question, which he then dismisses while proposing an eighth.[22] Allan first nominates meaning as ostension, meaning as use, meaning as a condition of truth, meaning as a function of stimulus/response mechanisms, meaning as concepts, meaning as abstract objects, and a naturalistic view of meaning. Each of these categories has difficulties for the author. Allan's own convoluted response to meaning appears not to answer the question directly, but through various linguistic steps that distinguish between "sense" and "denotation" he proposes that an expression corresponds to concepts in the mind of the user. But, in conformity with reductionism, these two are seen as different and apparently separate, which leaves us with a very confusing state of affairs.

Chomsky's view of meaning relates to the transformational relationship between the surface structure we observe (in Saussurian terms, the signifiers) and the "deep" semantic structure of the expression. At times Chomsky uses the term "sound-meaning" to denote this relationship,[23] or else he refers to this relationship in terms of psychology and grammar: "the deep structure is related to the surface structure by certain mental operations—in modern terminology, by grammatical transformations."[24] From a Chomskian perspective exactly what constitutes the "deep structure" is unclear, although Chomsky does say that "there must be a fixed system of generative principles that characterise and associate deep and surface structures in some definite way."[25] According to Roger Fowler, "the important aspects of the deep structure of a sentence are proposition and modality."[26] Proposition relates to a reference outside of language, and modality is concerned with the user's attitude to the propositional content of the sentence.

It is Chomsky's implied integration of deep and surface structures and his terminology of "deep structure" and "grammatical transformations" that I intend to appropriate into this study of meaning. Rather than describing the "vertical" relationship of language in terms of syntax and meaning that are linked by "grammatical transformations," I intend to argue that the process can be viewed in the reverse, and in a more integrated fashion. From such a position the deep, implicate, and symmetrical order of meaning is the semantic and psychological foundation from which all symmetry transformations are generated. These transformations construct as well as structure the infinite variety of forms on the surface, syntactic level of discourse. In simple terms this means that the three relations of meaning construct, by way of symmetry transformations, the signifiers and signifieds of discourse.

Finally, the philosopher Martin Heidegger identified the site of meaning not as some property of signs or objects, but as an essential of Dasein. (Dasein is a subjective and relational state that Heidegger describes by the compound expression, "Being-in-the-world.") "Meaning is an *existentiale* of Dasein, not a property attaching to entities, lying behind them, or floating somewhere as an 'intermediate domain.'"[27] This interactive site of Dasein, where the symmetrical structure of the system in general interacts with the reconstructed environment, is identified by Heidegger as the location of meaning.

Most of the writers I have referred to here view meaning simply as the single dimension of explicitness. But to discuss meaning from an isomorphic perspective we cannot simply accept explicitness as the only form of meaning. To reduce meaning to explicit meaning is to deny the asynordinate character of all semiotic activity and, more importantly, to ignore the unconscious and the implicate order of the universe. Yet such is the conventional and common view of meaning.

Not the Cogito

The distinction between primary and secondary meaning is isomorphic with the present approach to individual consciousness. I am therefore not referring to subjectivity in the secondary sense as a consciousness *of* something. This secondary sense is a consciousness as interpretation, as representation, as symbol, as signification that assumes a dualist subject/object formation. This is the sense in which consciousness is most commonly referred to by writers, as the process of bringing some objective or subjective state into conscious awareness. This predilection for the secondary processes of conscious awareness and representation will tell us little about the nature of the primary and structural conditions of consciousness.

As may be apparent by now, by referring to the structural conditions of consciousness I am not following Edmund Husserl's path of endeavoring to reflect on pure consciousness. The main difference between Husserl's phenomenological approach and the present one is that I am not trying to discover an essence of consciousness by reflective intuition. Rather, I intend to approach the problem through a study of meaning that will not provide an "essence" so much as the basic structure of the initial conditions of consciousness.

But what are the conditions of individual consciousness? Certainly not the usual cross-reference found in the subject indexes of many books—"*See* awareness." This view of consciousness traditionally equates a unified singularity of mind, soul, or Cartesian cogito with only an explicit conscious awareness. This exclusive approach to individual consciousness is usually content with the belief that conscious awareness operates on a rational basis excluding unconscious, "irrational" mental processes. These unconscious processes are commonly relegated to the body (and recently to neurophysiology), which is seen as separate from the mind.[28]

This Cartesian cogito of rational singularity is common to the cognitive-science assumptions that refer to "information," "computations," "coding" and "decoding," but is not the kind of subjectivity I am referring to here. Rather, individual consciousness as seen here refers to an asynordinate system that includes both unconscious and conscious states together, inseparable, interconnected. This interconnection is the nature of the asynordinate in that it involves both unconscious meanings that are implicit and conscious meanings that are explicit.

For present purposes there are five varying asynordinate conditions that make up the overall structure of individual subjectivity. These are

1. the unrepressed unconscious or, more positively, prereflective consciousness;
2. the Freudian notion of the repressed;
3. the Freudian preconscious where meanings are available for use;
4. reflective conscious awareness (Freud's W system); and
5. self-reflective conscious awareness.

These five states are used throughout this book to describe the distinctive features of individual subjectivity. It will be noted that these five states point to the changing balance of interaction between the implicit and explicit components of the asynordinate structure of individual consciousness. Primary meaning is dominant in the first three conditions, while secondary meaning has an increased role in the last two conditions. The overall relationship between the implicit and the explicit, that is, the unconscious and the conscious, has always been ambiguous and sometimes controversial. Freud seemed unable to decide on the nature of this relationship. However, I want to use a remark from Freud's *The Interpretation of Dreams* to exemplify the present isomorphic approach. Freud wrote, "The unconscious must be assumed to be the general basis of psychic life. The unconscious is the larger sphere, which includes within it the smaller sphere of the conscious. Everything conscious has an unconscious preliminary stage."[29] In this there are two asynordinate implications. First, we see that conscious awareness is a secondary stage in a process that begins with unconsciousness. In this sense conscious awareness does not stand alone as an independent or solitary agent but represents a secondary set of processes in a series of interrelated processes within the confines of the asynordinate structure of conscious/unconscious. Second, unconsciousness is *the* fundamental process of psychic life and as such has its own status. In addition, unlike conscious awareness, unconsciousness can stand alone as a solitary agent.

In other words, life can go on even when we are totally unconscious. In this particular instance Freud's description implies that conscious awareness is an emergent phenomenon. The asynordinate implications in this view contrast with the traditional Platonic and rational view that situates individual subjectivity as the free agent of mind or cogito that can transcend its own unconscious foundations. This instrumental view mistakes the part for the whole and takes conscious awareness as a separate and sole agent, divorced from its deeper unconscious structure.

Meaning as Consciousness

No one to my knowledge has seriously made the claim that meaning is separate or stands outside or beyond consciousness. To propose that meaning was unrelated to consciousness would be like speculating on the color of unicorns. Linguists and psychologists have generally assumed that there is a parallel or reflective relationship between these two but have been disinclined to ask more specific questions about its nature. For example, even though their discourse does not explicitly state it, linguists and psychologists (along with the general reading public) mostly assume that a conscious awareness is necessary to create and reproduce the meanings of texts when reading or writing.

This assumption of connection is also evident in the field of psychoanalysis. For example, when Freud wrote of latent and manifest content in dreams, he was referring to a content of meaning; he was not referring to a content without meaning. A dream without content is without meaning (not in the sense of a difficult interpretation but in the sense of no meaning) and would not register with the dreamer as a dream. According to Laplanche and Pontalis, the latent content of dreams is that "group of meanings revealed upon the completion of an analysis of a product of the unconscious."[30] They also describe the manifest content of dreams as the dream before it is analyzed, as it appears to the dreamer who recounts it: "By extension, we speak of the manifest content of any verbal product—from phantasies to literary works."[31] Freud himself writes directly about the meaning of dreams: "It is from these dream-thoughts and not from the dream's manifest content that we disentangle its meaning."[32] From these references there is no doubt that the psychoanalytic view is that both the spoken, conscious, and manifest content of dreams and the unspoken, unconscious, and latent content of dreams are contents of meaning. In other words, the use of the word "content" directly implies a content of meaning.

When Freud wrote of psychic acts, which are acts that can be conscious or unconscious, he gave examples of these acts in terms of "ideas," "images," or "presentations." Each of these terms is a conventional expression for "signs" or "symbols" or "representations," which are carriers and tokens of meaning. So basic is the proposal that the content of consciousness is meaning that it is hard to conceive of a psychoanalysis (the "talking cure") without also assuming this connection. Freud's approach to meaning, however, was traditional, for he never asked, "What is the nature of meaning?" but always "What kind of meaning does this or that psychic act manifest?" Thus it is in the *Interpretation of Dreams*. For psychoanalysis as for linguistics, meaning has always been a central but uninterrogated question.

Our difficulty in defining consciousness beyond broad categories and our confusion about meaning should therefore not be taken as the last word on these two phenomena. Further, I would suggest that this coincidence of indefinability is not due to chance but to the structural symmetry of meaning and consciousness. More precisely, consciousness and meaning are not parallel functions or processes that reflect on each other and somehow connect (the traditional view of linguists and psychologists). The relationship suggested here is much closer than that: a relationship of symmetry. It is proposed that (1) the processes and structures of meaning are identical with the processes and structures of consciousness; and (2) the processes and structure of consciousness are identical with the processes and structures of meaning.

Meaning's Structure

If consciousness and meaning share the same structure, then it is to meaning and the various relations of meaning we should turn to discover that common structure rather than to psychology. For many reasons related to the academic identity of the discipline, psychology has been unable to make many advances regarding the content of consciousness, and therefore its general mechanistic and reductive approach to subjectivity will not be of much assistance in this case.

In terms of subjectivity it will be suggested that the three basic relationships of symmetry, nonsymmetry, and asymmetry are realized within the general neurological mechanisms of excitation and inhibition, not in any specialized area but as a general function of the whole system, from reflex arcs to recursive cerebral cortical activity. The meaning of meaning therefore is not learned but is innately a product of basic and "normal" activity of a healthy nervous system. The formative, constitutive, semantic structure of meaning thus predates the acquisition of language and the interaction of signs, which are but secondary derivations.

Given these three basic relations of meaning, we can say that the structure of the various states of consciousness has the following general form:

1. Unconsciousness is produced by relations of symmetry. The Freudian concept of the repressed is dominated by the unconscious relations of symmetry that creates symmetry groups.
2. Prereflective subjectivity of sensations, feelings, and emotions is also dominated by symmetrical relations.
3. Preconscious states represent those processes that have been learned and retained for later recall. Recall processes operate automatically and unconsciously and involve the processes of selection and combination that in turn are structured by the series of relations of symmetry and nonsymmetry.
4. Conscious awareness is produced by a relatively complex, multilayered system in which nonsymmetrical relations predominate, resulting in attention, awareness, and, in humans, symbolic systems.
5. Self-reflective conscious awareness has the same structure as reflective awareness but with an inclusive organization and a focus on connection and integration.

As has been suggested, any constituent of conscious awareness, any point of attention, will always be an asymmetrical system. And such a system will always be constructed within the context of a series of larger asymmetrical systems (involving sensorimotor mechanisms as well as memory). Any so-called unit of consciousness (a single percept, concept or image) is therefore not some singular entity (like the fiction of a single unit of cognition or a single number or environmental object) but is always constituted as a secondary system of asymmetrical relations held in place by a complex series of larger asymmetrical formations that function mostly unconsciously.

What is incorrectly called a "unit of consciousness" is therefore a multilevel structure of interconnections, or in the words of chaos theory, a "fractal scaling" of systems, where the surface, macro constituent of attention is built up from the more general, deeper, micro systems that in turn are constructed from the primary relations of symmetry, nonsymmetry and asymmetry. The principal operation that ties together the various levels or scales of this integrated, multilevel fractal system of consciousness is that of recursive symmetry. Recursive symmetry links the whole multilevel system together, from the basic microlevel of the three relations to the macro constituents of awareness then to self-reflecting and identificational concepts that involve the self and the cosmos.

Given the recursive architecture of subjectivity and meaning, this architecture is made manifest at the point of reflective awareness and through some form of symbolic representation. This means that unconscious, symmetrical meaning will produce, by recursive generation, secondary meanings in the form of representations and signs. Further elaborations produce tertiary processes of signs relating to signs, or semiosis. Signs are therefore not seen here as Platonically separate, formal entities such as separate empirical or environmental objects. Rather, signs are constructed from the same subjective and recursive system of relationships that produces bodily sensations or the patterns of our environment. Signs are always constructed in discourse, which is the interactive system of individuals.

One further point needs to be made here in regard to discourse. It is assumed here that the discourses in which signs are constructed are not simply singular signals or linear "information." Discourse is an asynordinate complex of meanings rather than single lines on a page or information units in a computer. This view implies that all discourse, whether linguistic, scientific, artistic, dramatic, or performative, will be constructed by a variety of conscious and unconscious meanings. The structure of discourse, in other words, recursively reflects the asynordinate architecture of consciousness. This means that there will be more unconscious meaning in any sentence, painting, play, speech, mathematical formula or mime act than conscious meaning.

Unconscious meanings are those that could loosely be called implied, tacit, inferred, felt or sensed, and they represent the unconscious background context against which conscious, surface, symbolic meanings are construed. Unconscious, implicit meanings therefore make up the majority of meanings in any expression (and any system) as well as in life generally. In contrast, conscious meanings are those that are explicit and visible and come in the form of differentiated surface constituents. Conscious meaning is always the foregrounded meaning of constituent features and accounts for the minority of meanings in any expression, observation, image, or experience.

The Field of Meaning

The present isomorphic approach to consciousness proposes that all meaning is structured by relationships that have a sixfold set of purposes, referred to previously as formative, constitutive, structural, semantic, symbolic and immanent. Relationships, it will be recalled, are therefore not passive by-products of active mechanical systems, but are themselves active connective systems of energy, meaning and consciousness. Relationships are therefore always relationships of meaning whether they occur as the social interactions of particle or people or any other form of interactions. Meaning is thus the nonmaterial field constituted solely by relationships.

The field of meaning integrates what has traditionally been seen as the subjective inside, with the so-called objective outside as both are distinct aspects of the same field of consciousness. In this integrating sense, meaning pervades everything, and nothing can exist outside this field of meaning. In other words, meaning is always the primary field, which by its very nature supersedes all other "physical," mechanical, or dynamic fields discovered and established in physics, biology, mathematics, or any other discipline. Meaning is also prior to the organization of emotions, perceptions, observations, reflective consciousness, and symbolic representations, which are processes that emerge and are structured by the three primary relations of meaning.

One result of this priority of meaning is the acceptance that interactions between any of the myriad forms of the physical universe are relationships of meaning. These relations of meaning have all the qualities of cosmic consciousness. Such relationships as are seen to exist in the universe are always also present within subjectivity. The reconstruction of the world in subjectivity is only possible because of a commonality of category, that is, the category of relations of meaning. Subjectivity is thus able to isomorphically reproduce the forms, structures, and interactions of the cosmos only because of common relations of meaning. Like produces like, and so the cosmic relations, within layer 1, are thus reconstructible by layer 2, which is subjectivity, for subjectivity is already an emergent form from layer 1.

For example, particle interactions are relations of meaning at a quantum level. These we can isomorphically reproduce within subjectivity only because subjectivity is structured by the same relations of meaning that already structure particle interactions. If the interactions of particles were not relations of meaning, there would be two discreet or different categories of relations operating at cosmic and individual levels. If this were the case, particle interactions would have no meaning for us. That is, there would simply be a void of meaning in respect to particle interactions, and there would be no possibility of ever changing that void. In other words, we would be unable to refer to particle interactions in any way; there would be no representation of them, no symbolization, no mathematics, nothing. This is clearly not the case—quite the contrary, for the harmony and perfect symmetry of meaning, in respect to layer 1 and layer 2, is absolute. And there is absolute symmetry in respect to these layers of consciousness because they represent different aspects of a continuous, holistic field of meaning.

The logical effect of this isomorphic view is that we can never transcend meaning. As subjects we exist within this bounded yet infinite field, possessing observational processes that mimetically reconstruct our environment and having interpretive facilities that construe meaning through reflective and symbolic manipulation. Meaning represents the structural and psychological features of this cosmic and individual framework, and the formative relation of symmetry is the fundamental principle by which the whole structure operates.

This approach therefore assumes that the individual observer can be known scientifically, but not "objectively" as a biologist may want, nor introspectively as desired by some psychological approaches, but through a systemic model generated by the meaning of meaning. The meaning of meaning is that energetic field of symmetry which embeds each of us within its framework of cosmic consciousness, a field in which the physical world and all of its conditions are manifest, a field that operates personally through perceptions, sensations, feelings, and conceptions and culturally by way of discourse through reflective, symbolic representations.

As meaning is structured and construed out of the fundamental principle of symmetry, all the implications of this field arise from this principle. As I have argued, it is impossible to have meaning without consciousness or vice versa, so this field of meaning *is* also the field of consciousness. Consciousness therefore extends from the context of cosmic consciousness to individual subjectivity and to discourse. Each of these contexts of consciousness emerges from the preceding context, with cosmic consciousness the final and universal context. In each of these contexts the relations of symmetry recursively generate and transform self-similar formations of meaning at each level. The semantic transformations of discourse will therefore be the reflections of the transformations of prereflective subjectivity, which in turn will be the reflections of the transformations of cosmic existence.

Some General Characteristics of Discursive Meaning

This present isomorphic view of meaning does not attempt to develop a psychological theory by ignoring textual interpretation or vice versa. The universal *grammar* of meaning is applicable to deep unconscious meanings as to surface conscious ones. In terms of discourse it is therefore possible to make an isomorphic connection between Chomsky's surface and deep structure and conscious and unconscious meaning. Chomsky's surface syntax will always have some conscious elements, while his deep structure represents the implicate order of unconscious meaning. The universal grammar of the deep structure is therefore the formative grammar of symmetry potentials. The concept of a universal grammar of meaning is as applicable to deep psychological structures as to surface syntax or surface forms.

We can, therefore, describe the general characteristics of semiotic meaning in two forms: as conscious (explicit, surface, and differential) and unconscious (implicit, deep, integrated, and symmetrical). It should be remembered, however, that these two forms of semiotic meaning can never be separated in any representation. Together they represent the visible, which is held in place by the hidden.

According to Wittgenstein there are two conditions under which meaning can be said to function. These are (1) the verbal, the iterable, the symbolic, the what-can-be-said, and (2) the ineffable, the what-cannot-be-said. For Wittgenstein the ineffable is not *meaningless,* for its meaning can be shown. Showing and saying are not the same, however, for "What can be shown cannot be said."[33] The meanings of saying and showing are of a different order; one is expressible and the other inexpressible. I want to add to these two concepts of ineffable and expressible by relating them to implicate meaning (meaning that is unconscious and prereflective) and explicitly conscious and reflective meaning. In this sense "ineffable" meaning that can be shown but not said represents both those prereflective meanings of the body and those unconscious meanings inherent in symbolic representation. Secondly, the category of the expressible represents a combination of conscious and unconscious meaning in the form of signifiers.

It may be thought that the ineffable will always be predominantly implicit/unconscious and the expressible always asynordinate, that is, implicit and explicit. This is the situation for two conditions of meaning, but it does not hold for the third category of concepts and signifieds. We can see this more clearly by graphing Wittgenstein's two conditions of meaning with the other two categories.

In table 1 the ineffable, the what-cannot-be-said, has two types of operations. One is unconscious and comes in the form of prereflective bodily movements and sensations. In discourse these meanings involve the emergence of letters, words, and sentences through the subconscious processes of selection and combination. This activity would also involve the stimulation of sensorimotor processes associated with the movement of eyes, ears, throat, and tongue and other vocal areas when we are involved in "speech acts." The second type of ineffable operation is unlike the first in that it combines conscious and unconscious meaning in the form of concepts, what linguists call signifieds. Therefore, neither prereflective movement and sensations nor concepts can ever be expressed directly. Their expression is always indirect and through the representative agency of signifiers. The expressible therefore has only one set of operations, the symbolic, involving signifiers that like concepts are also constructed from both unconscious and conscious meaning. Signifiers come in the form of static units like letters of the alphabet or notes of music.

Wittgenstein's binary division between the "shown" and the "said" thus becomes more complicated when unconscious meaning is taken into account, for unconscious meaning represents a large part of both these categories. The integrating function of unconscious meaning also has the effect of subverting the binarism of the "shown" and the "said" so that there is an aspect of the shown in everything that is said.

The final claim I wish to make about semiotic meaning is that a sign as presented in this book is one that combines elements of physical, biological "givens" (the neurological activity involved in perception and conception) with elements of arbitrariness and cultural convention (the *type* of codes, symbols, and conventions used). These two areas have in the past been strictly separated, especially in the areas of structuralist and deconstructionist writing. The binary structure of Saussure's sign (signified over signifier) has tended to exclude any reference to the personal, the biological, or a material referent.

In order to describe the nature of the recursive terrain stretching from neurology to culture, I have added to Saussure's formalist sign structure (signified over signifier) the important foundational context of the body. This third contextual category of the body is combined with Saussure's signifier and signified to produce a new sign structure. This new sign is the *isomorphic sign*. The structure of the isomorphic sign provides a theoretical underpinning for all interpretation and discourse, a field that involves all human communication activities and can be thought of as the field of manifest consciousness. The three levels of the isomorphic sign (which represents a cross section of discourse in any moment of time) has the following "vertical" character:

3 Signifieds: conceptual structures.
2 Signifiers: perceptual structures.
1 Prereflective bodily meanings

These three levels of the isomorphic sign represent three dimensions of any cultural field. The fourth dimension is represented through a horizontal time line. This sign structure represents the invariant set of asymmetrical processes that, working together, produce discourse. This is an organizational structure of the consciousness within discourse. The vertical nature of this sign represents one slice of time in the production of discourse, one moment in speech, writing, reading, viewing a film, or interpreting the world in some way.

Table 1. Conditions of Meaning

	Implicit Meaning	**Implicit/Explicit Meaning**
Ineffable	(Prereflective bodily meanings)	(Concepts/signifieds)
Expressible		Symbolic (signifiers)

For example, when we read, we prereflectively reconstruct the individual linguistic units as signifiers within a frame of perception and also automatically construct the concepts that are produced by these signifiers.

The isomorphic sign has a double system of explicate orders that emerges from the implicate order of consciousness. This double system is not a straight-line, linear cause-and-effect process, from prereflection through to concept, but an emergent hierarchy, of system transformation developing from system transformation. For example, signifiers first emerge from the implicate movement of perception, then conceptual images (signifieds) emerge from the combinations of signifiers, and finally smaller concepts combine and transform into larger and more general concepts. This hierarchy of explicate levels represents a complexity of meanings: a complexity of systems generating further systems that generate further systems of meaning and so on.

This hierarchical complexity has several features that are similar to the turbulent character of dynamic systems in the so-called physical world. These features are usually confined to mathematics but appear to reflect the descriptive nature of the isomorphic sign. They are fixed points, periodicity, and chaos. For example, at the level of perceptual signifiers, fixed points emerge. This is the level of the fixed and unchanging signifier, the "key strokes" of discourse: letters of the alphabet, notes in music, phonemes in speech. In discourse the fixed points of signifiers moves (via recursive combination and selection) from single key strokes into the periodicity of word combination and sentence combination. Under some circumstances (such as anxiety or an uncontrolled reaction) this periodic phase in the production of discourse can develop into a chaos of meaning. A chaos of meaning does not necessarily emerge from a turbulent flow of words and sentences, as some chaos specialists might imagine—although for those individuals who experience logorrhea a chaos of meaning can and does come from this source. A chaos of meanings can also be contained within the symmetry of an emotional reaction (such as anger or guilt) that then stops or retards the periodic flow of meaning within sentence construction. A chaos of meaning can also emerge from discourse where there is a dominance of rhetoric over representation. In short, a chaos of meaning will emerge from the periodic flow of discourse when undifferentiated symmetry has become overly influential in any discursive act. (I also return to discussion in chapters 10 and 11.)

This fertile cross-referencing of the features of so-called physical systems of complexity and the complexity of meanings inherent in the isomorphic sign could lead to the establishment of modes of analyzing such diverse areas as literature, writing, mental illness, or forensic linguistics. Such an approach would need to take into account the different kinds of periodic flows in writing and in speech and the way in which some writing, through the use of metaphor and simile, tends toward the edge of chaos. These developments would only be possible, however, when there is a focus on meaning as a complex system. The common linguistic focus on "language," with its categories of grammar and lexicon, do not allow for these kinds of developments. This traditional focus only provides us with the possibility of variations on the theme of a Cartesian grid or network that recodes the codes and reclassifies the categories of "language."

As mentioned above, the time factor inherent in the isomorphic sign can be called the horizontal dimension of discourse. But it should not be considered as a single line, for it contains the three vertical levels of the sign operating simultaneously through a time dimension. This configuration, having both depth and length, is more like the four-dimensional field of space-time than the abstract model that Saussure has suggested for language. This model is in the form of a static cross that has vertical and horizontal axes. The present isomorphic sign does not have these linear features but represents a complex set of recursive processes that simultaneously work both vertically and horizontally in the dynamic production of discourse.

Finally, an important feature of the isomorphic sign is the emphasis on the context and specifically the context of the biological body. This is a biosemantic emphasis. We see this in the priority given the body in the first level of the sign. This valuing means that all communication, all texts, all so-called language will be constructed by a particular person(s) within a particular social and political context. This means that all communication will contain the political and social "interests" of its authors. In this sense, all discourse, even the discourses of science, is contextual several times over. It will contain the personal, cultural, and political interests of its authors. Therefore, no discourse exists without a context. It cannot be innocent and value free or paradigmatically free, for no discourse is constructed in a Platonic vacuum. From a isomorphic perspective all analysis of discourse, whether literary criticism, forensic linguistics, psycholinguistics, or some other form should always proceed by looking at the contexts in which the discourse is produced. The meaning of a context is the meaning of an implicate organizing framework from which explicit information emerges. Attention to contexts, both the critics' and the authors', is a way of interpreting texts so that the observer and the observed are contained within the single field of the interpretation. This is important in both "pure" and natural science, as it is in the humanities and the social sciences.

Meaning as Integration and Discrimination

Finally, we return to Benveniste's view of meaning, which lends itself to an isomorphic reading. Benveniste proposed that *meaning* must be defined in relation to *form,* and both these elements must be articulated within the whole range of language. For example, what constitutes the form of the linguistic unit is its capacity to be broken down into constituents of a lower level. In relation to *word,* the letter units that compose this word can be taken one by one: *w, o, r, d.* In this sense these are the formal units of the constituted *word.* These four letters separately, on their own, disclose the formal constituents, but their integration discloses what Benveniste calls "meaningful units." So it is that the phoneme, the smallest unit of speech, or the alphabet letter, the smallest unit of writing, are integrators (carriers of meaning) along with other phonemes and other letters, but they are also discriminators, units that have formal and constituent semantic potentials.

Benveniste tells us that meaning and form must be analyzed together, but the analysis for each proceeds from opposite directions so as to encounter them in the same linguistic entity. "Form and meaning thus appear as conjoined properties, giving of necessity and simultaneously, and inseparable in the functioning of a language."[34] Another image of the "conjoined properties" of meaning and form is given by Benveniste in a footnote where he refers to Saussure, who "conceived of 'meaning' as an internal component of linguistic form." Benveniste then quotes Saussure, who, in the original text of the *Course in General Linguistics,* refers to the structure of the sign. "The two-sided linguistic unit has often been compared with the human person, made up of the body and the soul. The comparison is hardly satisfactory. A better choice would be a chemical compound like water, a combination of hydrogen and oxygen; taken separately, neither element has any of the properties of water."[35]

While "form" and "meaning" as a pair of elements are different from the binary structure of the Saussurian sign, the metaphor of water used here by Benveniste appears to reflect the kind of transformative linkages common to systems and discourse. Saussure's structuralist view of language is one that sites meaning as a differential function of signifiers and signifieds: the two elements that combine to structure the sign. The role of difference is one that gives value to the linguistic units themselves as well as establishing the nature of the linguistic environment: how units (signs, signifiers, and signifieds) relate to each other. In this sense language is seen as a complicated network of differences that operate within the sign structure (between signifiers and signifieds) and also between signs. In this sense meaning is equated with difference. This view enabled Saussure to state that "in language there are only differences without positive terms."

Saussure's position on meaning is reductionist in that the formal constituent units of language are seen to have only a differential and distinctive character. In Benveniste's terms they are discriminators divorced from the integrating function of meaning. Yet Benveniste tells us that meaning has the function of integration, which is inseparable from the discriminating function of form. If meaning does have an integrating function, this function is all but neglected by Saussure. Saussure does state that both the signifier and the signified are "purely differential and negative" when they are considered separately, but that their combination "is a positive fact; it is even the sole type of fact that language has."[36] If this integration (which according to Benveniste is meaning) "is a positive fact," then Saussure is inclined to ignore it, for the overwhelming number of references in the *Course in General Linguistics* are concerned with the "negative facts" of language: difference. In other words, like the metaphor of water, Saussure's analysis, which focuses on one of the constituent elements of language, has tended to make language disappear in the manner that water disappears when we focus on one of its constituent elements. Perhaps this is why the *Course in General Linguistics* is not about language in use but is mainly concerned with language not in use.[37]

By using the metaphor of water in relation to the structure of meaning we can take Benveniste's description a step further. Thus meaning, like water, can be seen to have two constituent functions (integration and discrimination) that, when combined (at a certain level of complexity), operate as a linguistic identity, such as a phoneme or a grapheme. The metaphor of water suggests that these two elements of meaning are unified in the linguistic identity in the way that water is the unified result of oxygen and hydrogen. This means that where water is *constructed* from a transformation of oxygen and hydrogen the linguistic unit is similarly *constructed* from a transformation of the functions of integration and discrimination, of the relations of symmetry and nonsymmetry.

From the isomorphic perspective it is possible to take this analogy a step further by suggesting that the constituent functions (integration and discrimination) of meaning and therefore discourse can be analyzed separately from the linguistic unit (as oxygen and hydrogen can be analyzed separately from water). But when water is broken down into its constituent chemical elements of hydrogen and oxygen the unity of these two in the form of water then disappears. In the same way, if a linguistic identity can be broken down into its constituent elements of integration and discrimination then the unity of these two, in the form of a linguistic unit, also vanishes. In the first instance we have moved from perhaps the field of hydrodynamics to chemistry and in the second the transformation has taken us from the traditional field of linguistics to biosemantics.[38] This second change from linguistics to biosemantics is a central transformation and is represented by the interrelationship between the first and second levels of the isomorphic sign, that is, between the prereflective body and consciously perceived signifiers.

But the transformation from linguistics to biosemantics is not the only change that has taken place in this description. There has also been a sliding away from Benveniste's rather "objective" view of "form" and "meaning" into an isomorphic description of these two terms. The description concerns their functions of integration and discrimination, and these functions are spoken of as if they were functions of meaning. Our description indicates first that discrimination is a semantic function, one in which differential meaning, that is, differences are produced. Such a description conforms with the general stance of structuralist and poststructuralist thought that difference is the source and the effect of meaning. However, if we now go beyond the exclusive strictures of *only* difference as a constituent of meaning and as well accept Benveniste's position that integration is a producer of meaning then we will have two constituent functions of meanings: discrimination and integration. And when these two constituents of meaning are unified, as they must be, their unity produces the linguistic unit, as implied by Benveniste's approach.

But there appears to be a problem here. If discrimination (difference) is a constituent of meaning, then Benveniste's term "form" now seems to have dropped from the picture. Or has it? We see that Benveniste's description of "form" is one defined as the unit's capacity to be broken down into constituents of a lower order. But can the function of the unit be separated from the identity of the unit? In other words, is not the form of the linguistic unit its unitary structure, and does not the structure provide us with its identity, its form?

Jakobson's Metaphorization

If we were to accept this position, we would have settled on a view similar to Saussure's, one in which identity and function become welded into a unit's difference, which then becomes the formal and solitary meaning of signs. This view is well accepted in semiotic and linguistic literature, but it is a view rejected here. The ramifications of this traditional linguistic view need some elaboration. For example, this traditionally one-sided view is taken up and promoted by Greimas and Courtes in their criticism of Roman Jakobson's use of the concept of metaphorization. Greimas and Courtes argue that relations of meaning that function in language "are meaningful precisely only insofar as they create meaning, in other words, insofar as they create differences."[39] Apparently for Greimas and Courtes, as well as many others, meaning cannot be created by integration, or by the relations of symmetry, or its close cousin "similarity," each of which represents the dominant and integrating elements in metaphors.

This criticism of Jakobson centered around his assertion that certain aphasic conditions, such as defects in the "capacity of naming," were deficiencies on the vertical or paradigmatic axis of language. Other forms of aphasia, such as the "one-word sentence," were deficiencies of the horizontal or syntagmatic axis of language. Jakobson took the image of the vertical and horizontal axes of language (in the form of a cross) from Saussure. Saussure had distinguished two axes of language, a vertical one that he called the "the axis of simultaneities" and the horizontal axis, the "axis of succession."[40] The vertical axis of the cross is the dimension of language that various writers have claimed involves such functions as similarity, simultaneity, selection, association, and synchrony. It is also the axis of language pertinent to the construction of metaphors.

It should be noted here that each of these terms—symmetry, similarity, simultaneity, selection, association, and synchrony—implies the function of integration, the constituent of meaning to which Benveniste refers and of which Greimas and Courtes are so critical. Saussure designated the term *synchrony* as having the function of simultaneously gathering together linguistic elements into some form of hierarchy. Jakobson, moreover, informs us that "similarity connects a metaphorical term with the term for which it is substituted."[41] These descriptions of synchrony could be seen to represent a rough approximation of the internal relationships of a metaphor that operate on the basis of similarity. It is to be expected then that Jakobson found that the vertical axis of language can be represented by the notion of metaphor and that impairment in this area meant that the patient could grasp "the words in their literal meaning but could not be brought to understand the metaphoric character of the same words."[42] Such patients, however, apparently do have some functional ability on the horizontal axis of language, for they "answer with a metonymy when asked to repeat a given word, for instance, would say *glass* for *window* and *heaven* for *God*."[43]

Now in their criticism of metaphorization Greimas and Courtes suggest that Jakobson's exploitation of the dimension of resemblances (the vertical axis of metaphor and similarities) "amounts to abolishing meaning."[44] Such an extraordinary suggestion lies at the heart of reductionism in its framing of "language" as a network of differences. In contradiction to this stance I would argue that to suggest that similarities abolish meaning (as has also been suggested by Derrida)[45] is to indicate the reverse, that the meaning of similarity (and symmetry) is abolished. For linguistics this abolition of the primary and formative dimension of meaning means that similarity and substitution, which are the bases for metaphor (and metonymy for that matter), are dismissed as irrelevant. But such an view also suggests that the category of similarity, the general function of which informs the paradigmatic axis of language, is an illegitimate designation.

These "reasoned" responses of Greimas and Courtes as well as of Derrida are rejected here in favor of an affirmation of the integrating functions of similarity/symmetry that represent *the* radical constituent of all meaning, linguistic and nonlinguistic. Symmetry has an integrating function, the meaning of which inheres in terms like *association, synchronic,* and *reflexivity* and is structurally necessary in tropes like metaphor and metonymy.

It is suggested here that the vertical axis of language, which Jakobson has generalized by the term *similarity* (an axis, as Saussure pointed out, that excludes such differences as those of time) is structured by the relationship of symmetry and has the function of integration. This simply means that *all* the terms applied by different theorists to denote the character of the vertical axis of language will imply the relationship of symmetry and the function of integration.

The isomorphic view of linguistics locates symmetry as the fundamental and necessary constituent of all semiotic activity. The meaning of symmetry can be represented by the mathematical equals sign (=), a sign that is not meaningless but fundamental to all mathematics as well as to the issuance and production of all linguistic units and the structural transformation of these units into words, sentences, and texts. I therefore prefer Benveniste's more subtle and fertile descriptions of meaning to the structuralist and poststructuralist overdetermined belief in difference/discrimination.

From Benveniste's initial distinction of form and meaning we have progressed to the point where linguistic identity can be viewed in the manner in which we perceive water, in that its form "appears" to be present only in a transformed capacity: as a synthesis of two constituent functions, integration and discrimination, which represent an asymmetrical synthesis of the relations of symmetry and nonsymmetry (difference).

Appearance and Solidity

The linguistic unit, from the present perspective, therefore represents a three-element structure: the two constituent elements of meaning, symmetry, and difference (having the functions of integration and discrimination) and the third element that is their unity (the asymmetrical synthesis) of the other two and that forms the linguistic unit. From this integrating, semantic view the basic relations of meaning, when synthesized together, transforms from the dynamics of relationships into the "solidity" of a linguistic form that "appears" in conscious awareness as a linguistic presence or identity.

The phrase "appears to be present" is critical to the above description. It is critical because it destabilizes the common linguistic view of a "form" as a stable object "out there" that we call a linguistic unit or identity. If we think of a linguistic unit as a material form that is objectively present, existing "out there on the page" divorced from human intervention, as Saussure saw *langue,* then the phrase "appears to be present" is disruptive. Yet it is this disruption to the "objective" perspective that is necessary when dealing with the underlying semantic transformations that construct linguistic units. While our "objective" view of water can accommodate its transformation from two combustible chemicals into a noncombustible liquid, it seems more difficult to view a linguistic unit as a similar transformation of two semantic constituent functions. Why is this the case? The answer seems to be a intellectual tradition that has denied a place for meaning and consciousness in relation to observing the world.

A linguistic entity such as the letter *a* or the numeral *1* is a transformation that takes place not on the page or on the blackboard, not "objectively" out there, or formally in a transcendent zone, but within a field of meaning that involves a cultural history, physical stimulations, the activation of eyes or ears, and the higher nervous centers in the brain. In other words, linguistic entities, like all entities, are constructed within the sensory field of the individual by the individual. Such constructions are transformations involving the conscious and unconscious functions of recognition, identification and memory that in turn produce a mimetic reconstruction of the letter *a* or the numeral *1.* The phrase "appears to be present" now has some relevance to the above discussion, for the "presence" of the linguistic unit, seen from this position, represents not a "solid" material object "out there" or a formal entity, but a series of simultaneous events linked together within a field of meaning. The identity of a linguistic unit is therefore the unity of a system constructed in consciousness.

This viewpoint does not mean that the "solidity" of the linguistic unit is a fiction, or an illusion that can be replaced by a more solid reality. It does mean, however, that the apparent solidity of a linguistic unit or a text is an habitual construction of conscious awareness that functions almost entirely unconsciously. As symmetry is the implicate order of this field of meaning the emergence of an explicit "object," in this case the letter *a* or the numeral *1* will be a symmetrical construction, in other words, a mimetic construction. A mimetic construction is one in which a physical field has interfered with a perceptual field to produces a single "packet" of meaning, a unity, a mimetic construction.

The seemingly contradictory view suggested here reflects, but is not synonymous with, the Nietzschean view of truth as a kind of error.[46] However, the truth of a linguistic unit is only an error (of reason) if we entertain the hope, like Derrida, that we can do without the "fiction" of its presence. In other words, if we believe it is possible to speak and write without using the meaning of integration, of symmetry (which of course *is* impossible, as linguistic units are only formed because of the integration of differences into a unified whole). As it is not possible to construct linguistic units from only negation/difference, the truth and reality of a linguistic unit is of the transformational kind we find in any identity. Since transformations are necessary if there is to be consciousness, the transformations that construct linguistic identities are as "real" and as "truthful" as meaning and consciousness. The truth of the mimetic construction of a linguistic unit is therefore a necessary truth if we accept that consciousness and meaning are necessary for the construction of "reality."

The distance travelled in this description from Benveniste's linguistic view is considerable. Not only has the site of the linguistic unit changed from "objectively out there" to a field of meaning with mimetic constructions, but, in addition, the nature of meaning has changed from a kind of formal "propositional function" (as Benveniste suggests)[47] to a set of semantic relations that are produced by an individual's nervous system and that give rise to subjectivity.

Jakobson's study of different types of aphasia supports this suggested shift of position from the "objectively" out there to reposition the linguistic unit as a transformation brought about by the activities of an individual consciousness. The various conditions of aphasia are conditions of the individual's semantic and neurological behavior, and this view forces us to resite the phenomenon of "language," a resiting from Saussure's transcendental *la langue* to the constructed field of meaning involving behavioral processes of communication. John Searle also rejects the formalism of transcendental language when he speaks of the production or issuance of a symbol, word, or sentence.

> [T]he unit of linguistic communication is not, as has generally been supposed, the symbol, word or sentence, or even the token of the symbol, word or sentence, but rather the production or issuance of the symbol or word or sentence in the performance of the speech act.[48]

The prereflective context that constructs the "presence" of the linguistic unit in the performance of a speech act implies that linguistic units themselves will have the added characteristics of that context. In other words, a linguistic unit such as a phoneme or a letter of the alphabet or a number will not only have an embedded meaning that discloses integration and discrimination, but will itself represent aspects of consciousness and unconsciousness. In other words, it will be asynordinate.

It is recognized that linguistic units differ one from another. Phonemes fit with systems of speech, while letters fit within written discourse and numbers fit within mathematics. Yet while these systems of discourses are distinct, their emergence within subjectivity follows similar transformative paths. I will have more to say about speech and writing in following chapters. As for numbers, this discursive system displays some interesting features.

Unlike the units of speech and writing, numbers can be communicated individually or combined in any fashion. In addition, numbers appear to closely represent the structure and function of the primary relations of consciousness, symmetry, nonsymmetry, and asymmetry. For example, the zero can be seen to represent generalized symmetry, the implicate order or the nonexplicit. The number one is the first unit or explicit "packet" of meaning and contains the functions of integration and discrimination that are the values of the relations of symmetry, nonsymmetry, and asymmetry. As a unit of explication within conscious awareness, the semantic values embedded within "1" can be extended in a linear mode both forward or backward or reflexively iterated in a loop for as long as a mathematician has the patience to perform symmetry transformations. From this isomorphic perspective mathematics can be seen as a direct reflection of the three semantic relations of consciousness. Perhaps for this reason Platonists have often seen mathematics, like the cogito, as divine.

Conclusion

In the past meaning has been a mystery to both linguists and psychologists. Yet meaning is always assumed to be present in all semiotic and psychological activity. Cartesian assumptions about meaning and consciousness have tended to see both as visible, explicit, and differential. The present isomorphic view argues that meaning is constituted by the primary relations of symmetry, nonsymmetry, and asymmetry, and these generate secondary systems of meaning that inhere in all material forms as well as in symbolic representation. These three formative relations give rise to the asynordinate structure of both meaning and consciousness, for these states are identical. Bohm's explicate order is therefore an order of hidden and unconscious meaning, and this is the foundation of both cosmic consciousness and individual subjectivity. In similar manner the explicate order is made up of differential meaning that has the double function of realizing the material objects of the universe and having them realized in conscious awareness. It was also argued that an isomorphic sign should replace Saussure's binary sign in any science of signs and that some linguistic accounts of meaning (Benveniste) lend themselves to an isomorphic reading.

Chapter 6

Freud and Matte-Blanco

In this chapter I want to begin to show that subjectivity is not a featureless unity but has a well-defined semantic structure made up of the three relations of symmetry, nonsymmetry, and asymmetry. The question addressed here is similar to that which Noam Chomsky has called a "universal grammar." Chomsky has said that "the study of universal grammar, so understood, is a study of the nature of human intellectual capacities."[1] From the present perspective the three relations of meaning are a "universal grammar" that recursively constructs cosmic and individual consciousness in its latent and manifest forms.

I will begin to deal with the evidence that suggests the structural content of consciousness by looking at the Freudian view of consciousness and unconsciousness. In this regard I will be drawing on Freud's own work as well as the writing of Ignacio Matte-Blanco, who has proposed that the unconscious functions by way of a symmetrical logic, while the conscious mind operates on the basis of an asymmetrical logic.

The Psychoanalytic View

What evidence is there to support the contention that the relations of symmetry represent the structural content of unconsciousness and nonsymmetry and asymmetry the structural contents of conscious awareness? The evidence used here is twofold: in the first place, it comes from Freud's view of consciousness and unconsciousness and from the way his view has been transcribed by Matte-Blanco; secondly, the evidence comes directly from the internal logic of *symmetry, nonsymmetry,* and *asymmetry* and how the meaning of these terms relate to what we know about unconsciousness and consciousness.

For Freud (and almost every other theorist who has written after him)[2] the psychic system of conscious awareness is largely indefinable. Freud bracketed together conscious awareness and perception (he called it "the W system") and saw the function of this system as the ability to receive sensible and differential qualities from the external world. But this is hardly a description of awareness. Freud tended to separate the "perception-consciousness system" from other psychic systems where memory traces can be inscribed (such as the preconscious and the unconscious). Overall, however, the transient property of conscious awareness that traditionally is seen to make distinctions and discriminations about the world at large is left by Freud as a featureless product of our intuition. As Laplanche and Pontalis say:

> Freud holds consciousness to be a fact of individual experience lying open to immediate intuition, and he makes no attempt to define it beyond this. It is "a fact without parallel, which defies all explanation or description. . . . Nevertheless, if anyone speaks of consciousness we know immediately and from our most personal experience what is meant by it."[3]

In response to this lack of detailed description of awareness Laplanche and Pontalis call for a theory of consciousness that makes its function and position clear. Apart from the present study there have been some attempts at what could be called a theory of consciousness. Matte-Blanco, writing in 1975, put forward some suggestion about the features of this so-called featureless state and so began to address the problems of consciousness. He proposed what he saw as the essential ingredients of consciousness. He tells us that "consciousness could not exist without asymmetrical relations, because the essence of consciousness is to distinguish and to differentiate."[4] In this proposal he is referring to Freud's W system of conscious awareness and not to the general category of consciousness that includes unconsciousness.

In a project designed to reformulate the Freudian system of consciousness and unconsciousness, Matte-Blanco views the mind as dynamic—having impulses, instincts, drives, wishes, and desires that may be in conflict with each other—but as also having the properties of a mathematical set. Mathematical set theory is concerned with the items and their relationships within a given set or collection. Matte-Blanco states that his writing "is an attempt at a clear formulation of the unconscious . . . in terms of some logico-mathematical notions."[5] The main thrust of his writing is concerned with the features of the unconscious and thus with a symmetry set. But true to the binarism of a traditional rationalist, Matte-Blanco opposes symmetry to an asymmetrical set, and therefore unconsciousness is placed in opposition to conscious awareness.

As we have seen in chapter 4, asymmetrical relations are those kinds that are irreversible and that operate by hierarchically linking at least two relations together, as in "Peg is the mother of Amanda," "Paris is east of London," "A is part of B." For Matte-Blanco, conscious awareness has an asymmetrical structure: "the mode of being revealed in consciousness . . . requires, as an essential feature, the use of asymmetrical relations. Consciousness could not exist without asymmetrical relations, because the essence of consciousness is to distinguish and to differentiate."[6] Yet while Matte-Blanco views conscious awareness, as Freud did, as "spatiotemporal," in contrast to the unconscious,[7] he nevertheless believes that we cannot focus on more than one thing at a time. Thus, when confronted by a class of items or objects, our conscious awareness "can only consider it in two ways: either it focuses on the limits (or definition) of the class, that is, on those precise features that characterise it and distinguish it from all other classes, or it must conform to its nature, which prevents it from seeing more than one thing at a time."[8]

This is a rational approach that wants to control each explicit detail in order to end up with the linear description of "one thing at a time." Matte-Blanco does not explain how a system of asymmetrical relations (involving an integrated series of relations) can at the same time have a focus on "one thing at a time." The two definitions conflict, for the "one thing at a time" definition implies only linear explicitness, while an asymmetrical set of relations implies a parallel set of relationships that are simultaneously present in various stages of awareness. This conflict between one thing at a time and a system of integrated relations represents the classical conflict between a rational and an asymmetrical paradigm.

From an isomorphic perspective, conscious awareness has an asymmetrical and an asynordinate structure. Such a structure always has a background of implicitness and a foreground of explicitness, in other words, varying levels of conscious awareness. In order to satisfactorily arrive at this integrated and complex system containing varying degrees of awareness, it is necessary to introduce a third term to Matte-Blanco's binary approach. This third term is of significance for the philosophy of language. It is nonsymmetry or difference.

We could say that a focus on "one thing at a time" is a reified, rational focus on difference. But the term *difference* does not constitute a definition of conscious awareness or, for that matter, of the concepts of system. The concept of difference does, however, provide the explicit constituent of any system of asymmetrical relations. Thus if the explicit constituents of any system are differences, these nonsymmetrical relations will not stand alone but are integrated aspects within the overall context of a system of relations. This means that the processes of integrating differences is another way of saying that the variety of items in conscious awareness at any one time will possess varying degrees of awareness. Therefore, when Matte-Blanco states that "the essence of consciousness is to distinguish and to differentiate," we can interpret him to mean that nonsymmetry is an essential feature of the asymmetry of conscious awareness, or that differences will always be the conscious constituents of conscious awareness.

Read from the perspective of reason, Matte-Blanco's examples of asymmetry always emphasize the differences of a system and tend to erase the symmetries of system context. But we do not have to focus on the differences of a system; we can instead focus on the unconscious integration of the differences and thereby the recursive context of the system. From the isomorphic perspective of this study asymmetrical relations have two features: integration and differentiation, which are combined (by symmetry) so as to contextualize individual differences into an overall system. Read in this manner, the basis of consciousness as a system is the complex interaction of the relations of symmetry and nonsymmetry, that is, a set of connected differences that make for a differential but self-referencing system.

For Matte-Blanco, who worked from the perspective of reason, asymmetrical relations are those that distinguish and differentiate. That was the extent of his description, except perhaps insofar as these relations are seen also as the outward expressions of a conscious Being. From an isomorphic perspective, asymmetrical relations represent both the outward expressions of a conscious awareness and the internal structuring of that state.

But the generality of Matte-Blanco's proposal is accepted here. The basis of this agreement is that conscious awareness has an asymmetrical structure. However, from an isomorphic perspective I will redefine his proposal: conscious awareness has the semantic structure of asymmetry. This system provides for the integration of constituent differences into a semantic and a psychological context.

The difference between these two views is due to the differences between Matte-Blanco's traditional perspective, which stresses the exclusion of differences, and the present isomorphic perspective, which focuses on the recursive integration of differences. Matte-Blanco's binary approach is seen in such divisions as "The Two Modes of Being in Man," and the "logic" of symmetrical and asymmetrical relations is presented as "bi-logic." Added to this dualism is the predisposition to see entities in place of processes. Matte-Blanco also views awareness from a Cartesian perspective, as almost a "pure" and absolute state, untainted by the symmetry of unconsciousness. Conscious awareness and unconsciousness are therefore seen as almost separate entities.

In his introduction to *Thinking, Feeling, and Being,* David Tucker interprets

Matte-Blanco's view of consciousness very much in these exclusive terms. He informs us that "a modicum of symmetry plays a part in *any* discrimination of a set, or, in other words, in any normal classificatory act by the mind."[9] From this point of view conscious mind is perhaps almost pure, having only a modicum of unconsciousness. These views contrast with the present one, which views any act of conscious awareness not as pure state but as the result of an emergent process involving more unconscious than conscious activity.

Like Freud, Matte-Blanco concentrates mainly on the unconscious. He emphasizes Freud's distinction between the repressed and the unrepressed unconscious and sees unconsciousness as having several degrees of depth. Overall, however, Matte-Blanco equates the relation of symmetry with unconsciousness and tells us that "throughout this book I shall use the expression 'the unconscious' extensively, understood in the sense of symmetrical being."[10] He also runs the two terms together, as in the "symmetrical unconscious."[11] Thus, from his perspective, the functioning mode of the unconscious (both repressed and unrepressed) is one in which symmetrical relations are produced. The logic of the unconscious is therefore symmetrical logic: a logic that treats the converse of any relation as identical with it.

From the psychoanalytic point of view this means that asymmetrical relations (past histories for instance) can be treated by patients symmetrically (past events occur in the present without an apparent time differential). Symmetrical logic is therefore a loose description used by Matte-Blanco for what he sees as the logical operations of the unconscious mind that function on the basis of the principle of symmetry.

This connection of symmetry with unconsciousness has the following character. All the unconscious defense mechanisms described by Freud have at bottom the common unifying principle of symmetry. For instance, the mechanisms of projection, sublimation, transference, and the return of the repressed are each seen by Matte-Blanco as examples of displacement. And displacement functions on the basis of symmetry, which is a principle whereby two or more separate objects are treated as one. Using an example of a patient who treats his professor as his mother, Matte-Blanco says: "When displacing (for instance, from mother to professor) the unconscious does not treat both only as possessors of something in common, but in fact treats them as identical."[12] Thus when a professor takes on the attributes of a mother for a student, the two figures are not perceived by the patient as having something in common but in the relation *mother equals professor equals mother.*

By the same argument Matte-Blanco sees the fundamental mechanism of condensation as also functioning like displacement, that is, on the basis of symmetrical logic. Condensation is usually seen by Freudians as an unconscious process whereby different associated meanings accumulate at significant "nodal points" of meaning or around certain images that, especially in dreams, acquire exceptional intensity.[13] Matte-Blanco suggests that the unification of meaning accomplished by condensation is possible only through the evidence of symmetrical logic.

In addition to these defense mechanisms, several other characteristics of the Freudian unconscious are exemplified by symmetrical logic. These are the absence of negation and the absence of contradiction. Freud informs us that the basis of the unconscious consists of wishful impulses that are exempt from mutual contradiction. "There are in this system no negation, no doubt, no degrees of certainty."[14] In the same paper Freud also tells us that the processes of the unconscious are timeless, that "they are not ordered temporally, they are not altered by the passage of time; they have no reference to time at all."[15] Each of these aspects of the unconscious is viewed by Matte-Blanco as evidence of the principle of symmetry; thus there can be no contradiction, no negation, no degrees of certainty, and no dimension of time in symmetrical relations. From Freud's writing Matte-Blanco identifies thirteen characteristics of the unconscious and finds that there is one element common to each: "that they unite or unify things which for ordinary thinking are distinct or separate."[16] Thus for Matte-Blanco, while ordinary conscious thinking "usually works within a framework of distinguishing things, the unconscious that Freud investigates tends to unite and fuse everything."[17] The conclusion that Matte-Blanco reaches therefore, is that the unconscious works through the principle of symmetry.

An Isomorphic View of the Unconscious

In his reformulations of Freud's conscious and unconscious, Matte-Blanco is clearly not concerned with the problems of language, or with meaning. He writing is influenced by Cartesian traditions brought to bear on a psychoanalytical perspective and seen in unmediated terms of *mind, being, thinking,* or the rational/mathematical terms of *sets* and the *logic* of these sets. He does not write about discourse, semantic relations, or the relations of meaning that are my concerns here. Viewing Matte-Blanco's Freudian reformulations from the present isomorphic perception, we conceive the relationships of symmetry and asymmetry, not as abstract mathematical sets, but as relationships of meaning. Everything suggested by our argument so far indicates this banal but important point; that here are relations of meaning that structure consciousness and discourse rather than some abstract, formal, Platonic mathematical category of a set.

However, Matte-Blanco's creative proposition that the unconscious is the symmetrical mode of being is in general accepted here. My isomorphic emphasis is only slightly different. The ontological terms of "Being" or "bi-logic" and the notion of a mathematical set are not used in this book. Instead I describe the connection of the unconscious and symmetry in the following manner: *unconscious activity functions on the basis of symmetrical relations, and the relations of symmetry produce the quality of unconsciousness.*

The reversed symmetry in the above description is never explicitly stated by Matte-Blanco; the closest he comes to it is this: "The interesting thing about it [the symmetrical mode] is that the quality of being unconscious is a *necessary consequence and only in this indirect sense* a constitutive feature of symmetrical being."[18] I consider the description above a much clearer one because no foreign framework imposes an antagonistic set of values on the isomorphic connections being described. This tendency for crossing paradigmatic boundaries is symptomatic of instrumentalism and reductionism when used to describe, in linear terms, some form of isomorphic relationship.

A further change in the present perspective is the integration of awareness and unconsciousness. These two are always integrated asynordinately, but not in Matte-Blanco's argument. One of the central questions to be asked by anyone writing about consciousness is how separate awareness is from unconsciousness. If these are exclusive states, we will tend to view unconsciousness in the traditional psychoanalytic sense, as contrary to "ordinary thinking," marked perhaps by a display of aberrant behavior. Conscious awareness will therefore be seen as Matte-Blanco describes it, as a kind of "ordinary thinking" in which distinctions are made.

These divisions raise a difficulty with Matte-Blanco's reconstruction of Freud's view of conscious and unconscious states. I agree with Matte-Blanco's general points that "consciousness could not exist without asymmetrical relations"[19] and that the features of the unconscious are seen in the "application of the principle of symmetry."[20] Yet Matte-Blanco consistently writes about a "symmetrical-asymmetrical duality."[21] In other words, the problem is not the major proposition of his work (that asymmetry equals awareness and symmetry equals unconsciousness) but the manner in which these two "modes of being" or agencies supposedly interact.

If, as Matte-Blanco assumes, conscious awareness is relatively separate from unconsciousness, we will speak unhesitatingly about "ordinary language" or "ordinary thinking" as separate and different from the thinking, and presumably the language, of the unconscious. Matte-Blanco states that "it is not clear whether there is a continuity from conscious to unconscious. It seems more likely that it is rather a question of an 'either or.'"[22] But if this duality holds such that psychiatrists and psychologists can make rough assessment about norms and neurotic behavior, it cannot hold in relation to meaning and discourse. If this duality held for discourse, we would be unable to write statements that contained the meaning of unity, identity, symmetry, or similarities or even statements about isomorphic patterns. We should not be able to do this because each of these terms embodies and represents unconscious states (symmetry). On the basis of the exclusion of unconsciousness from conscious awareness, we should not even be able to know about the unconscious.[23]

If this traditional and exclusive view of consciousness holds up, then statements about symmetry, unity, similarities, and totalities should, on the basis of the separation of awareness from unconsciousness, be meaningless to awareness or at least have no known conscious meaning. Now this is patently not so, for we routinely make statements about a whole variety of symmetrical relationships (which according to Matte-Blanco and the present writer are unconscious). How is this possible?

I suggest that it is possible because conscious and unconscious meanings are never separate. They are integrated and operate asynordinately. In other words, statements about symmetries are made by a language system that is always an asynordinate and reflective system. Every statement will therefore be constructed by the inherent symmetries of this reflective system. Because the three relations of meaning subtend and structure any statement, the meaning of symmetry will never be foreign to us in any statement. The linguistic effect of this integration is that in every linguistic unit (whether letter, word, sentence, paragraph, or book) there is an asynordinate structure, that is, a composite process of conscious/unconscious meaning.

If this is the case, then every statement, whether it is about symmetry or difference, will be understandable to us because it will be isomorphic with the structure of consciousness that produced it. Because of this common structure, statements about unconsciousness and symmetrical relations will have meaning for us (even if, as we shall see, they cannot be *known* consciously, visually, and differentially in any pure sense). In other words, human beings do not have a decoding mechanism that operates purely consciously and explicitly (as assumed by cognitive science). This could be the case only if conscious awareness were separate from unconsciousness. When we are involved in the process of reading or writing, that is, producing texts and discourses, the complex processes of recognition, identification, recall, and linguistic choice operate asynordinately: both unconsciously and consciously, synthesized together in a series of complex processes.[24]

Before I describe the evidence for the integration of conscious and unconscious qualities, it should be noted that Freud positioned himself ambiguously in regard to the separateness of conscious awareness and unconsciousness. If we ignore Freud's conflicting topographical and economic evidence and accept his position in *The Interpretation of Dreams* that "everything conscious has an unconscious preliminary stage,"[25] then this implies a process of integration. On the other hand, from Freud's functional position the transient conscious stands in opposition to unconscious and preconscious systems of memory trace, and therefore a separation of the two is implied. In addition, if we accept the Freudian position on the question of "knowing" the unconscious, this also implies a separation, for the unconscious, according to Freud, can be known only indirectly, through the elliptical language of dreams, jokes, slips of the tongue, et cetera. As Freud says: "Unconscious processes only become cognisable by us under the conditions of dreaming or neurosis."[26] Thus the Freudian view on *knowing* is that we can consciously know the unconscious only indirectly, by reading the latent meaning from what is manifest in disrupted signs, nodal points, emotional images, and dreams. In the sense that the unconscious is filled with latent meanings, the concepts of meaning and knowing were separated by Freud in a manner consonant with his rational approach and his functional view.

The present isomorphic emphasis reworks this question of the relationship of conscious to unconscious meaning by stressing the evidence of asynordinate integration. The evidence is overwhelming. We may not know (in a pure and rational sense of consciously, visually, and differentially knowing) the relations of symmetry. We may not be able to differentially prove the existence of a symmetrical relation, but that does not imply that the unconscious meaning of symmetry is a void, meaningless nothing. It simply indicates that the meaning of symmetry is not one recognized by the secondary system of conscious awareness.

Implications of Integration

Matte-Blanco's proposed two modes of being imply a division within subjectivity. However, the separateness implied by his two modes of being does not live up to its assumed binary status, for it is contradicted by the very language used to define their relationship: "Perhaps it is equally true to say that without symmetry there could not be asymmetrical functioning in man. Asymmetrical relations are something that emerge from and come out of the sea of symmetry."[27] This interesting statement squares well with Freud's comment in *The Interpretation of Dreams* that the unconscious must be assumed to be the general basis of psychological life, it being the larger sphere including within it the smaller sphere of the conscious.[28]

These statements by Freud and Matte-Blanco undermine the traditional binary perspective that is instrumental in both their writings. This is not some wayward forgetfulness on their part, but a natural outcome of focusing on the general isomorphic character of "relationships" and in particular the relationship between conscious and unconscious activity. This is the case because relationships will always foreground the function of integration, a function produced by the relationship of symmetry and inherent in terms like *relationship*. In contrast, a Cartesian and binary perspective foregrounds differences (the nonsymmetry of oppositions) that provide the focus for separateness and division.

The force for the integration of awareness and unconsciousness can be seen in the very terms Matte-Blanco chooses to use to describe their structural features. *Symmetry* and *asymmetry* carry strong implications of integration. For example, a set of asymmetrical relations implies a twofold integration: (1) a set of symmetrical relations that come prior to this set, and (2) a set of symmetrical relations that forms part of the asymmetrical set. The term *asymmetrical* therefore implies the kind of integrated process that both Freud and Matte-Blanco have referred to in the last two quotations. And these are references in which the asymmetry of conscious awareness represents a secondary process that is constructed out of an unconscious "sea of symmetry." The written difference between these two terms rests on the prefix *a-*, which, unlike the letter in Derrida's *differance*, is an audible sound. The similarities of the terms come from the rest of the word. These terms suggest overlapping processes of integration and isomorphic patterns rather than distinctions that produce a binary separation.

In addition, if it is accepted that the structural content of the unconscious is symmetry and the structural content of conscious awareness is asymmetrical relations, then the two quotations above appear to be saying the same thing, namely that unconscious symmetrical relations are the basis of psychological life and that the asymmetrical relations of conscious awareness are contained within that larger sphere. Yet while both Freud and Matte-Blanco at various stages saw the integration of unconsciousness with conscious awareness, the binary framework from which they both working determined the organization of their material in such a way that they were forced into dualistic interpretation of these essentially integrated processes.

The other factor that most probably influenced Freud and others concerning the general separateness of conscious from unconscious activity is the mechanical imperative to classify behavior into the duality of either sane or aberrant. Such common binary classifications have affected Matte-Blanco's terms of "ordinary language" versus unconscious thinking. This neat division, which still has legitimacy in psychiatry and the law, is destabilized by the asynordinate character of subjectivity. This integration would appear to mean the end of any clear-cut division between rational, sane behavior and nonrational or irrational, aberrant behavior. The integration of conscious and unconscious meaning appears to call for a new set of definitions of "mental illness," sanity, insanity, and unsanity. Such a large project, which would be based on an isomorphic paradigm, is perhaps now possible but extends beyond the scope of this book.

The Implicit as the Implicit

Matte-Blanco tell us that to answer the question "why the Freudian unconscious is unconscious is relatively easy if it refers to the repressed unconscious."[29] The content of the repressed unconscious may become conscious once the barrier of repression has been removed (through therapy), but, as Matte-Blanco is at pains to point out, the unrepressed unconscious because of its very nature cannot enter consciousness.

The concept of the unrepressed unconscious represents that which cannot become conscious. And that which cannot become conscious is structurally distinct from the repression. While Matte-Blanco has much to say about the unrepressed unconscious, he does not detail the various positive forms symmetry can take. In contrast to Matte-Blanco, I see the symmetry of the unrepressed unconscious as having positive and appropriate forms:

1. Semantic prereflectiveness involving bodily activities
2. The "preconscious" involving learning and memory
3. A major part of conscious reflective awareness
4. A major part of self-conscious reflective awareness

Each of these forms and conditions differs from the irrational of symmetricization, but only relatively. While the contents of repression may become conscious, the preconscious mechanisms that select and combine memories will always operate unconsciously and so in a manner that will never become conscious. We therefore do not have a binary division between the repressed and the unrepressed, or between irrationalism and rationalism, or between the unconscious and conscious awareness. Rather, because of the asynordinate structure of subjectivity there is a series of conditions of awareness that contain varying degrees of unconsciousness.

For Matte-Blanco the answer to the question "What is it about the relation of symmetry that is unconscious?" represents a kind of mystery. He accepts that symmetry cannot be known consciously and suggests that our unconscious thinking functions in higher numbers of dimensions than our conscious thinking.[30] While this is an intriguing conclusion, it is based again on a binary separation of these two states, and for this reason is rejected here.

I turn to another example to demonstrate the seeming contradiction that the relation of symmetry cannot be known consciously even though it does have meaning for us. First, what is a symmetrical relation? As we have seen, symmetrical relations can be said to exist when X equals Y. Implicit in this equation is a reverse symmetry, Y also equals X. In the term *spouse,* for example, or the statement "Fred is married to Sam," we see a reversed or circular relationship operating. Traditionally, the results of symmetrical equations have been interpreted as the state of identity, with the reversible equation written as A = A. Now in each of these examples I have limited the description of symmetry to simple equations. By doing so I have also reduced the conscious and differential content of the expression (the signifiers) in order to try and leave only the unconscious meaning. But here is the problem. If I do away with all conscious content (surface signifiers), I will have departed the field of expression and so will not be able to demonstrate anything. Yet by using surface signifiers (which are always constructed consciously and differentially), we can never capture directly and explicitly the character of symmetry (unconsciousness).

How to express the inexpressible? The closest we can come to it is to say that the meaning of the equal sign (=), automatically and involuntarily felt or sensed by mathematically literate people, reproduces the character of symmetry. The meaning of symmetry, therefore, does not reside with the explicit, differential marks of the signifiers, the writing we "see" on the page. Rather, it resides in the background contexts out of which the explicit signifiers have emerged. If this description appears too vague, then let us remember that the meaning of the equal sign is central to all mathematical calculations.

In order to prosecute this argument let us analyze the equation A = A. In this equation, where do we find the physical evidence, mark, or feature of symmetry? The first A is similar in form to the second A, but it is different from the second A in space and time. In addition, the two bars that signify the equal sign are also similar but again different in space and time from each other and different again in shape and form from the two A's. So where is the symmetry? All we have consciously found here are marks that are different from each other.

It is true that we have also found similarities, but *similarity* is not symmetry. If symmetry represents the structure of unconsciousness and difference (nonsymmetry) is the mark of conscious meaning, then a "similarity" will be the dialectical result of the two, of symmetry and nonsymmetry, a consciousness that is both conscious and unconscious but oriented toward the unconscious of integration. That which is similar implies a link or commonness of at least two things. In other words, for the first A to be similar to the second A the two units need to be separate while at the same time to have something in common, that is, their symmetry, in this case the symmetry of their form. "Similarity" is therefore the direct, dialectical result of the integration of symmetry and nonsymmetry. Terms that also convey this dialectic, such as *resemblance, likeness, analogy, metaphor, substitution, correspondence, combination*, and so on, represent the same kind of asynordinate balance between differentially unconscious meaning and integrated differential meaning.

However, our problem of unconscious meaning is not with similarity but directly with symmetry: unconscious meaning. Where then do we find the physical evidence of symmetry? The answer appears to be that there is no physical, visual, and explicit evidence in these or any other signifiers. This is because signifiers (the surface and second level of the isomorphic sign) will always appear only in conscious awareness as marks of difference. In other words, Freud's perceptual-consciousness system is characterized by differences. Freud assigned to the psychic system he called the perceptual-consciousness system (the system W) the ability to receive sensible qualities. In the words of Laplanche and Pontalis, "consciousness gives us what are called qualities—sensations that are different in a great multiplicity of ways and whose difference is distinguished according to its relation with the external world."[31]

Now I suggest that in order for signifiers to function as signifiers they have to be consciously distinguished "according to their relation with the external world" and each other. And this relation is one of nonsymmetry (difference). The rational approach that Freud and others employ in regard to consciousness automatically establishes a system of exclusive differences (nonsymmetries). This focus comes from a reinforcing and circular reflection on the differential marks of awareness. In other words, a differential awareness of differences, or a rational focus on the foundations of reason, produces a system of exclusive differences.

The recursive logic of this means that there never can be any conscious and perceptual evidence of symmetry. Quoting Freud, Laplanche and Pontalis tell us that "within this difference there are series, similarities and so on, but there are in fact no quantities of it."[32] Freud could just as well have said that while similarities come within the bounds of conscious awareness, the underlying relation of symmetry does not. A similar view is held by Matte-Blanco, who states that "it must be clearly understood that symmetrical being is not conscious and that, on account of the nature of man's consciousness cannot ever become conscious."[33]

From both Freud and Matte-Blanco, as well as from the argument presented here, we see that the unconscious meaning (of symmetry) cannot be known in conscious awareness. Rather, it represents the ineffable foundation of subjectivity, cosmic consciousness as well as discourse. Thus it represents the ineffaceable foundation of all meaning, and therefore symmetry will always have meaning for us.

Unconsciousness Formations

I now wish to return to the point concerning the unconscious nature of symmetry and to explore this in more detail. The intellectual traditions of the explicit are so dominant in Western thinking that to write about the character of the implicit is often to invoke disbelief. Skeptical predispositions aside, however, some comment is required for the five types of unconscious formations that structure every subjectivity. These five are symmetricization, or what has traditionally been called the irrational; mythological discourses; and three types of closure: necessary, relative, and absolute. Each of these are discussed below.

Symmetricization

One of the main characteristics of the unconscious, identified by Matte-Blanco, is "symmetricization," a term he uses instead of the irrational. In many respects symmetricization is a better word because it implies the symmetry structure of unconsciousness, whereas the irrational refers to one part of a binary pair that implies nothing about structure. Symmetricization is a functional term that refers to the irrational, undifferentiated logic of unconscious mechanisms like condensation and displacement. Undifferentiated logic comes from a fusion of meanings, meanings that are compressed by a process of compound identification. In *Thinking, Feeling, and Being,* an example of irrational thought is cited. In this case a particular breast belonging to a particular mother was symmetricized by a patient (her son) into the concept of *breastness* encompassing all breasts of all women.[34]

The kind of compound identification in the logic of symmetricization can be represented by a symmetry group, $A + B + C = X$, in which, for example, A, B, and C could represent a range of features in a personal history while the X equates to a present emotional "trigger." The fusion of each of these individual items represents the compound closure or identification. Symmetricization in this form can be seen as a series of ego-defensive reactions.

Mythological Discourse

The fusion of meanings seen in the concept of "breastness" is not only an affect of a personal idiosyncratic unconscious, but this kind of fusion of meanings is also common to myths, for example, in stereotyping, naturalization, unqualified categorizations, and overgeneralization. In other words, I am suggesting, first, that symmetricization is a compound fusion of meanings that are seen to be irrational because of their often arbitrary combination in mechanisms like condensation and displacement and, second, that such symmetricizations are also common to mythological and prejudicial discourses. In this recursive sense, mythological discourse will reflect the traditional Freudian mechanisms of the repressed.

Symmetricization is also the structure of mythological propositions and naturalizing discourses. An example of naturalizing discourse relevant to this discussion would be a statement that "objective reality" is made up of natural objects like tables and chairs that are separate from ourselves. Here we have a fusion of the following: (1) the use of particular syntax and grammar (2) the concepts that emerge from the expression (3) the particular environmental condition referred to (4) the structure of subjectivity. All of these distinct features are identified together in the construction of "objective reality." This kind of compound identification is common in the natural sciences, in which unmediated "reality" is taken as given. (Mythological discourses are discussed further in chapter 12.)

Propositional contents that are symmetricized in discourse and thought (silent speech) are never completely unconscious, for there is always some element of nonsymmetry, of conscious awareness. If there were no element of conscious awareness, there would be total symmetry and therefore total unconsciousness. Thus the system of thinking and expression in symmetricization is of the type in which the relation of symmetry is the *single* most dominant structural value. This is the structural implication of the term symmetricization. As a framework for subjectivity, symmetricization will not lead to a detailed analysis of anything, for its organization of meaning is dominated by the compound closure of its constituent parts. Distinctions and differences are therefore not made in situations that deal directly with systems of distinctions. In this closed sense, a symmetricized perspective constructs its own self-justification, either through its reactive and uncontrolled behavior or, in terms of discourse, through the unconditional imperative of its expression. Mythological discourses therefore have the same symmetricized basis as defense mechanisms and often serve the same purpose of defending the ego of the author.

Closure

Another general type of unconscious formation that structures and organizes subjectivity is closure. In general, a closure is a completion or a conclusion by a closing off of implicit meaning. From the isomorphic perspective closure is found in every aspect of consciousness, subjectivity and discourse. Closures are formed when explicate unities emerge from their implicate background contexts and in these emergent processes implicit meaning is closed off or excluded from the formed unity. In *The Role of the Reader* Umberto Eco writes about "open" and "closed" texts. According to Eco, open texts tend to encourage acts of conscious freedom, while closed texts aim at arousing a precise response from the reader.[35] In contrast to Eco's rather singular interpretation, I suggest that closure is not a special element of a particular realist narrative style. Rather, closure is a general and widespread structural feature that operates recursively throughout consciousness, subjectivity, and at every level of discourse.

The closing off of implicit meaning occurs when unities emerge in the cosmos as physical entities. At some point the cosmic symmetry of the background is closed off in order that a distinct, explicit entity can emerge from its background. If this closing off did not occur, there would be no emergence and so no explicate order, no physical world, and also no subjectivity with a conscious awareness. The closures that produce physical entities (rephrased in Pauli's exclusion principle) are mimetically reproduced in subjectivity through the same processes of emergence operating as conscious awareness. This set of emergent processes also exists as discourse when linguistic units, such as key strokes (letters of the alphabet and numbers) and words are mimetically reproduced within subjectivity. Closure can be said to be an unconscious process that constructs units and unities, the meaning of which is the basis for differentially "knowing" the world.[36]

Closure can be considered as an enfolding or "packaging" of meaning. Packaging is a double semantic process. By this I mean the enfolding of meaning within a "packet" that appears as a unit or unity so that the unit both contains hidden meaning and at the same time displays visible, explicit meaning related to the surface distinction of the unit. For example, a unit of the English alphabet, like the letter *a,* or of mathematics, like the number *1,* encloses within itself the symmetrical meaning associated with the unit's construction. At the same time the process of enfolding creates the unit surface form by which we know it as distinct and different from other similar units. Any unit of information, such as alphabet, number, musical note, or phoneme will therefore possess internal symmetrical meaning that is hidden and external differential and surface meaning that are the distinctions of conscious awareness.

In the philosophy of Aristotle and his concept of identity, expressed by the equation A = A, we find an early rationalist attempt at coming to terms with the closure of meaning. This equation displays a closure through the equal sign that forms the unity of the identity in which the first A appears identical to the second A. The resultant enclosing of these two constituents features into a single entity is then distinguished from similar entities as well as against a background context of nonidentity. By this logic Aristotle developed his rational "Laws of Thought."

Necessary Closures

What is significant about the closure of units, whether in discourse (as signifiers) or in quantum mechanics as particles, is that the units need to maintain the invariance of their identity by maintaining their differences. For difference to exist there has to be the closing off, in a specific location, of the background symmetry. This closing off is achieved by the principle of emergence (see chap. 3), in which two sets of symmetries interact and produce a set of nonsymmetries. This is just another way of referring to the Pauli principle. In other words, semiotic units maintain their identity by being necessarily closed *to* the vicissitudes of critical reflection as well as closed off *from* their implicate background. In this manner they can be used as the necessary mechanical building blocks of sentence and concept construction. This kind of necessary invariance in semiotic systems reflects the necessary invariance of Pauli's exclusion principle, which leads to the building-block development of life from its inorganic foundation. The development of discourse is therefore reflective of the development of life.

It should be remembered that semiotic closure always occurs at the level of prereflective consciousness. If a grapheme or phoneme were open to reflective interpretation, it would not be closed and therefore would not maintain its differences and so would be impossible to communicate. It would be like having a bodily reflex action open to reflective interpretation, or like being able to observe the very processes of observation while observing. In the manner that all prereflective bodily behavior is based on the invariance of involuntary reflexes, so communication and observation are based on closed and invariant meaning. In the case of communication, the involuntary reflexes are the invariance involving the production of linguistic units. (It should be noted that sentence construction is also based on the *relative* closure of syntax and vocabulary.)

The concept of a necessary closure does not prohibit any specific closed unit becoming an interpretative system. In other words, any grapheme of phoneme or simple environmental object can become the point of conscious reflection and thereby be transformed into a recontextualized system having internal components. Most of science is about undertaking this very process. Such conceptual reflection does not invalidate the category of the prereflective and the "necessarily closed." While one unit (of discourse or the physical world) may become the focus of attention, to be transformed into an interpretative system, this process of transformation is being carried out by a reflective consciousness that is also using the closed building-blocks of a prereflective consciousness in order to undertake this process. In linguistic terms this means that such an interrogation will always be carried out by the involuntary use of linguistic units that are themselves closed to interpretation during their prereflective use. The category of the "necessarily closed" is therefore not related to a specific unit but to the applied functions of prereflective consciousness.

Closures do not always have two equally balanced values (like Aristotle's concept of identity) that are united and fused together. Take, for instance, the sentence "The car is around the back." Focusing on the concrete noun "car," we can say that it involves a closure formed by two very different sets of features. Here we have a closure involving *(a)* the word "car" united with *(b)* prereflective, nonexpressible, complex of meanings from perception and memory. The two are quite unequal in the sense that the word is different from our sensations of the concrete materiality of the world. Yet here we have, because of the way concrete nouns "work," a closure constructed from the "presence" of perception (past and immediate) and the representation of a noun. This closure of ineffable experience with symbol represents the structural meaning of a concrete noun. Concrete nouns would be unrepresentable if this necessary closure did not occur.

Relative Closures

Some closures are also relative. For example, closure operates conceptually in the relationship between a particular and a class. In this symmetrical relationship there is no suggestion of a completely reversible equality to the extent that the class becomes the particular. There is, rather, a connecting relationship; for instance, a rose is a flower but a flower is particularized, not generalized, by the term *rose*. The two terms *rose* and *flower* are nonsymmetrical (different) in regard to their irreversibility, yet they share a symmetrical relationship that produces their class connection to each other: their context. This connection is their commonality, represented in terms like *generalization* and *particularization*. There is not a total closure or fusion of the two constituents into one, as there is with the meaning of identity. Rather, there is a relative or contingent closure, brought about by the sharing of a common meaning, or more precisely, a sharing of a symmetrical relationship. Relative closures have constituents that are connected yet different, symmetrical yet nonsymmetrical, conscious yet unconscious, integrated yet distinct, open yet closed.

Perhaps the most central necessary closure for philosophy is found in the notion of truth. From an isomorphic perspective truth is like the evolutionary term *fitness:* a successful organism "fits" the conditions of its environment. In this sense of fitness, truth is to be found where there is a closure of conscious awareness regarding several elements. For instance, truth emerges where the grammatical rhetoric of a sentence fits together coherently and is internally consistent; or where a set of propositions and theories fit known evidence accumulated by prereflective sense impressions; or where, through conventional use, a particular word becomes a noun or a verb and fits the object or action referred to; finally and most significantly, when the reality of the world is involved, where we perceive the world about us and the percept fits, and is symmetrical with, our processes of perception. In each of these cases—consistency, correspondence, physical reality—a truth is constructed. In each of these cases there is a closure of differential conscious awareness because the "fitness" of two or more constituents has resulted in a closure of meaning and a formation of a truth.

In this sense truth is not the singular result of rational deduction arrived at logically and through propositional forms. Rather, truth can represent a complex of closures of conscious awareness that permeate the whole of subjectivity at every level of discourse from alphabet to sentence, from signified to referent, both rhetorically and representationally. Truth is therefore not a singularity but a web of closures operating throughout every discourse and at every level of subjectivity. The important question is therefore not a traditional binary one of truth or falsity that relates to single propositions, but what kind of truths are operating in what contexts and for what kinds of purposes? This is a contextual and a political question.

Absolute Closures

Not all closures are relative, and therefore not all truths are relative. The closure of the field of meaning is an absolute closure in that nothing exists outside meaning. Another way of saying the same thing is to refer to the recursive architecture of subjectivity. In a holistic sense the symmetrically recursive structure of subjectivity reflects holographically the symmetrically recursive structure of the cosmos. Each subjectivity will therefore contain within it the same isomorphic patterns as cosmic consciousness. The world is in us and we are in the world. A process theologian like John Cobb would say, "God in us and we in God." The holistic closure between us and God is therefore an absolute one. In this sense the closure of the structure of meaning and consciousness is an absolute closure and therefore an absolute truth.

But in a contradictory manner, typical of an isomorphic perspective, truth can be both relative and absolute depending upon the context. What does this mean? It means that the bias of this study is in favor of isomorphic perspectives to the degree that our orientation coincides with the structural reality proposed. In other words, the "should" and the "is" of perception are the same.

I am therefore making the age-old claim that we should use a particular view of the world because it is the correct one, that an isomorphic perception is a perception in tune with the structural reality of perception. This is the kind of thing Hegel said about reason, that "the real is the rational and the rational is the real." The danger of making such a claim about an isomorphic paradigm is that, like Hegel, it could be wrong. If this is a danger, it should at least be foregrounded. But I do not see it as the logical end point of beginning with the concept of symmetry. I take this question of truth up again in chapter 11.

In summary, there are five types of unconscious formations that structure subjectivity. These are symmetricized meaning, mythological discourses, necessary closures, relative closures, and absolute closures. Symmetricized meaning has traditionally been called irrational (thinking and behavior) and is constructed by formation of symmetry groups of arbitrary features. Mythological discourses achieve the same fusion of meaning within discourse. Necessary closures are associated with the mechanics of building, building discourses and building the physical universe. Relative closures and truths are concerned with conceptual unities and are always open to interpretation and therefore they are only provisionally closed. Absolute closures are those concerned with the recursive and symmetrical structure of reality, hence the finite closure of infinite meaning, hence the identification of "God in us and we in God."

Conclusion

From this discussion we have seen that some Freudian texts on psychic states lend credence to an isomorphic interpretation. Thus the symmetry of the unrepressed unconsciousness can take several forms. Meanings can be fused together to form symmetricized formations that can be either in the form of defense mechanisms or mythological discourses. In the past psychoanalysts have been interested in symmetricized meaning only as an indication of personal aberration. From the present discussion we see that in addition to so-called irrational behavior there are also necessary, relative and absolute forms of unconsciousness. These are the necessary, relative, and absolute forms of closure and truth.

Chapter 7

Paradigms of Consciousness

The next step in examining the semantic structure of subjectivity is the secondary relations of nonsymmetry. These are the key relationships in the structure of conscious awareness. Conscious awareness is usually assumed, by rationalists, to be the "mind"—the featureless unity of a Cartesian cogito. Such assumptions run counter to the present approach. As all studies and discussions of consciousness are framed within the sight lines of some underlying perspective, these sight lines will predetermine the kind of conclusions we draw about consciousness. Therefore, at this juncture it is important look at some of the underlying contextual assumptions of the present isomorphic approach. In doing so, we will be exploring the concept of paradigms of consciousness.

Paradigms

What is a "paradigm of consciousness"? A paradigm of consciousness refers to a predisposing framework of subjectivity, and not to a methodological paradigm or a set of institutional practices as proposed by Thomas Kuhn.

Ever since Kuhn wrote *The Structure of Scientific Revolutions* in the early 1960s, his concept of a scientific paradigm has been controversial. Kuhn proposed that science did not develop through a steady accumulation of knowledge gathered objectively through innocently detached observation. Rather, the history of science has shown relatively long, stable, and uncontroversial periods of endeavor that are then transformed, by intellectual revolutions. According to Kuhn, these revolutions are produced by the relativity of scientific knowledge, by the existing paradigm or worldview of science being challenged by a new paradigm. The most obvious examples of a change in paradigm accompanying scientific revolutions are associated with the names Copernicus, Newton, Lavoisier, and Einstein.[1]

While there has been some debate about the term *scientific paradigm,* in general use it refers to a disciplinary framework of coherent principles and theories. Kuhn, addressing the question directly, suggests that it represents the relative unanimity of professional judgments found in a particular community of specialists.[2] This emphasis highlights Kuhn's notion of "tacit" knowledge, the shared knowledge that is accepted as a given by the specialist community and passed on to novices in the processes of learning and induction. In this sense, becoming conversant with the principles and theories of a scientific paradigm represents a form of scientific literacy that, once learned, will function preconsciously in the trained scientist as inscribed habits.

The concept of a paradigm of science implies that scientific knowledge is not independent of human construction and cultural convention. In other words, science is dependent on particular disciplines that have social, institutional, historical, and conceptual contexts. The concept of a paradigm puts in question our belief in raw data, neutral facts, and objective viewpoints. Rather than standing as independently existing phenomena "out there," data, facts, and viewpoint are determined by the paradigm, for the nature of a paradigm is to construct, mediate, and give meaning to these phenomena.

In addition, because the concept of a paradigm socializes and historicizes science, it also implies the involvement of consciousness. A paradigm refers to a worldview, an organizing perspective toward the material being studied. For example, when we compare paradigms we will alter the way we perceive the world. To perceive the world in a "classical" scientific way has come to mean that we should accept that the world is like a machine, open to deductive and mechanical computations. In contrast, to perceive reality from a holistic or isomorphic perspective is to see it as an undivided universe.

Kuhn, the historian of science, paid little attention to consciousness, and I believe it was this omission that helped create his problems of definition. In a postscript to the second edition of *The Structure of Scientific Revolutions* Kuhn admits that one reader has found at least twenty-two different ways in which he has used the term *paradigm*. His answer to this diversity is to say that a paradigm is broader than a theory and that it involves a disciplinary matrix of shared language, values, expectations, and beliefs. These further elaborations of definition have not clarified the term. Implicit in such descriptions is the assumption that subjectivity operates in each of these elements—language, values, expectations, and beliefs.

Rather than ignoring subjectivity, this should be our prime focus. Such a foregrounding of subjectivity will then lead to a distinction between a paradigm of consciousness and a paradigm of application, that is, of methods and practices. This distinction would mean that a *paradigm of consciousness* refers to the perspective framework, a framework of consciousness, and not directly to the methodological framework or set of institutional practices emphasized by Kuhn. It should be stressed, however, that it is impossible to separate perception from application, for our consciousness of a situation will always be determined to some degree by our practices and habits of involvement, and our practices in turn will be guided by our perspective framework. The two paradigms therefore reinforce and overlap each other. For example, insofar as the calculations for measuring linear systems are different from those measuring nonlinear systems, the difference is one not only of mathematical method but of a change of perspective, of subjectivity.

This overlap of paradigms of method and perspective represents a dialectical relationship implying, not their separation, but the integration of their differences. This integration persists even when solutions to scientific problems cannot be calculated using existing methods. For instance, the tension derived from a failure of a method to solve a puzzle does not lead directly to a new paradigm of consciousness. N. Katherine Hayles tells a story about the early difficulties of calculating the gravitational fields of the moon, the earth, and the sun. She tells us that "the problem was sufficiently complex and interesting to merit a name and a prize of its own. It became known as the three-body problem, and the king of Sweden offered a reward to the first person who could prove a solution was possible."[3]

In 1890 Henri Poincaré published a paper showing that a solution was not possible using Newtonian equations. Poincaré's paper implied that a new kind of mathematics was necessary to account for the dynamics of complex systems. The effect of this difficulty was to help define the limits of the existing Newtonian paradigm of method, but it did not suddenly produce a new worldview. A scientific revolution is brought about over time by changes in methods that challenge old worldviews sufficiently to allow the simultaneous construction of new worldviews and new methods. The process is a dialectical one in which method and viewpoint may reinforce each other at times of development and stability and resist each other during times of paradigmatic crisis. Paradigms of application and consciousness should therefore be seen as distinct but interconnected. They should not be seen as separate formations. In other words, the same set of distinctions can be perceived differently from two different subjective perspectives.

The distinction between a paradigm of application and a paradigm of consciousness, however, needs to be made clearly. A paradigm of application is usually developmental and specific to history, culture, and social setting. In contrast, a paradigm of consciousness is concerned with the organization of subjectivity, with the semantic principles by which any particular knowledge is organized. Paradigms of consciousness are therefore not developmental, nor are they specific to history or a particular culture. They are generic to all social settings and to all social and cultural interactions for the whole of history.

Yet these two paradigms are also interconnected processes. Relativity and absolutism are therefore combined in any and all knowledge. Thus the relativity of specific mechanisms of application, which are constrained by their cultural and historical context, are always synthesized with the semantically absolute of a particular subjective perspective. It is possible to explore this dialectical relationship between a paradigm of application and consciousness in the abstract terms of system *types* (a paradigm of application) and a paradigm of the *system in general.* This exploration could be developed by way of those features of application, such as different methods, contents, and literacies, which can assist or hinder a particular generic paradigm of consciousness. But here we would need to detour into the whole field of learning theory and transformational pedagogy, which for the present study seems too great a diversion. For present purposes it is sufficient to highlight the isomorphic (integrated yet distinct) nature of these two paradigms.

The Three Paradigms of Consciousness

One of the main proposals of this book is that the structure of meaning is simultaneously the structure of the whole architecture of subjectivity. This means that the integration and interconnection of different features of subjectivity (unconscious memories, awareness, recall, sensory stimulation, etc.) is achieved through the common, formative, semantic relations of symmetry, nonsymmetry, and asymmetry. In addition to proposing a three-part structure of meaning and consciousness, I also contend that each of these three relations can and does function as an orientation point for conscious awareness.

Conscious awareness is therefore never a simple searchlight of awareness that innocently and objectively records sense impressions. First it is an emergent, secondary system, and then it is also an interpretative system that always contains a predisposition or orientation. The orientation point of conscious awareness provides the interpretative structure and framework of disposition that orders and organizes identified and recognized impressions. In other words, recognition and identification are habitual and emergent processes that are part of a larger system of awareness that organized them into interpretative systems. An interpretative system is an interpretative perspective, or what is called here a paradigm of consciousness.

An orientation point represents a locus, or basic reference, a privileged and base semantic value or coordinate from which the perspective is built and organized. The orientation of a paradigm can also be called a "priority coordinate" in a perspective system of interlocking coordinates (the paradigm). In the three-relation system of subjectivity, there are three and only three possible priority coordinates and, as a consequence, only three possible paradigms of consciousness. It follows that the three elements of meaning—symmetry, nonsymmetry, and asymmetry—provide the orientation, the priority coordinate for each of the three interpretative framework of paradigms of consciousness.

If like Bohm and Peat we are concerned with the notion of order, it will mean that a paradigm of consciousness provides the ordering and the organizational structure for any particular set of impressions, practice, or forms of application. Again, from the point of view of order there can be only three kinds of orders by which all knowledge is structured and therefore perceived and expressed. I have called these three perspective frameworks paradigms of consciousness, for they are formative frameworks that order and determine the organization of all knowledge and practice.

The proposal for three paradigms of consciousness creates a distinction between organization and structure. The underlying structure of any form of consciousness will always remain invariant, for it will always have the same structure of symmetry, nonsymmetry, and asymmetry, but because of the formative character of these three relations the organization and values that emerge out of this invariant structure can be varied to form three orders or three paradigms of consciousness. In other words, the same structure can be organized in three different ways depending upon the basic value or orientation of the paradigm. The three paradigms of consciousness I call *symmetricization, reason,* and *isomorphism.*[4]

As suggested, each of these three organizations of consciousness is determined by its orientation. But in turn the orientation is characterized by the nature of its recursive patterns. For example, symmetricization, which has an orientation of symmetry, will have a pattern of recursive symmetry that produces compound identification. On the other hand, a paradigm of reason will have recursive patterns of nonsymmetry (differences), which create distinctions that are exclusive, which when taken to extremes produce dissociation and fragmentation of knowledge, institutions, and sensibilities. Again, an isomorphic paradigm will have recursive patterns of asymmetry leading to the integration of differences that when taken to extremes produces the integration of the individual with cosmic consciousness or God.

The orientation of a paradigm is therefore produced from the type and the intensity of its recursive patterns. As there are three types of paradigms of consciousness, consequently there are three kinds of recursive patterns. In the case of a symmetricized paradigm the intensity of repetitive symmetry can be widespread and acute, as when there is a series of compound identifications, in which case there will be a fusion of many associated meanings, and these will provide a symmetricized perspective. In a symmetricized paradigm the world is ordered into a series of symmetrical groups, in which all the meanings within any group are equal and have the same value. A reaction to the world based upon a symmetricized perspective tends to be overgeneralized, reactive, and highly emotive.

For a paradigm of reason there will be recursive patterns of nonsymmetry. Another way of thinking of the orientation of reason is in terms of a circular and reflexive set of differences, a closed differential reflexivity; in other words, differences that lead to more differences that lead to more differences and so on. In a paradigm of reason, the world is ordered differentially, so much so that the highest value for any individual to aspire is to fragment knowledge by drawing finer and finer distinctions.

In contrast, the type of reflexive pattern found in an isomorphic paradigm is that of recursiveness itself, that is, the recursiveness of integrating patterns, for example, of context and constituent that are ordered recursively. The structural key to this paradigm is transformation. All transformations entail an asynordinate change from hidden implicit meaning into explicit and visible meanings. Thus every constituent can be transformed into a context containing further constituents, while every context will generate a series of constituents. This recursive pattern of transformations (a symmetry transformation) from constituent to context to constituent can be extended infinitely until it has included the infinite context of cosmic consciousness, or God. The orientation of an isomorphic perspective therefore generates recursive transformations that can produce a profusion of recursive complexities at various levels or scales.

The difference between a paradigm of reason and isomorphism is sometimes not appreciated. The difference is to be found in their organization, their different kinds of reflexive orders. A consciousness of reason has an exclusive and differential reflexivity, while an isomorphic perspective has a recursive and integrative reflexivity. These different orders and organizations need to be appreciated; otherwise, isomorphism will be viewed as just another kind of reason.

These three paradigms of consciousness can be seen to be a very rough approximation to Merleau-Ponty's three kinds of knowing.[5] However, because they represent the three interrelated organizational fields, these three paradigms are more connected and formative than Merleau-Ponty's "unconscious drives," "pure ideas," and "being-in-the-world."

A summary of the main characteristics of these three paradigms of consciousness follows.

Symmetricization

Symmetry represents the general structural base of the implicate order of subjectivity, yet when this relation, through intensive recursiveness, becomes the dominant value in the generation of symbolic meaning, it creates the problems of symmetricization.[6] Symmetricization represents the double enfoldment of experienced or symbolized meaning that needs unfolding through some process of ventilation, such as articulation or reexperiencing. Symmetricization can result in large, fixed, and inflexible groups or "blocks" of meaning that are characterized by so-called irrational reactions, prejudice, oversimplification, and gross generalization. Symmetricization is the basis of neurotic and psychotic behavior.

However, symmetricized meaning is not constrained by the category of "mental illness." The orientation of symmetricization is an intensified recursive symmetry that means an unconsciousness of unconsciousness. This double unconsciousness can be seen as being unaware of what we do not know, or failing to name our own weaknesses, or being unaware of the effects of our behavior on surrounding people, or not being in touch with our unconscious. These are common reactions in young adults and also for many of us who have grown older.

The organization inherent in symmetricization is one of arbitrary and undifferentiated singularity, where, for example, an A can equal B which equals C which equals D, and so on indefinitely. This is an undifferentiated linking of meanings in a circular singularity (a symmetry group). The fusion of this kind of meaning is evident in what Freud called the repressed unconscious, with its wide range of defense mechanisms, where anything can arbitrarily equal anything else. This is the sense in which Matte-Blanco used the term symmetricization, as an aberrant enfolding of repressed meaning.

Freud's discussion of instincts and their vicissitudes is an example of how symmetricized meaning operates. Yet while Freud divided instincts into two groups, self-preservation and sexual instincts, in general much of his interpretation of behavior reflects a symmetricized perspective. That is, Freud himself employs a symmetricized perspective in his interpretations of instincts, for a good deal of his interpretations reduced behavior to the ways sexual instincts (the libido) can be expressed. This form of crude reductionism, which is common to much of mechanical psychology, represents a symmetricized perspective in operation, that is, an unawareness of the limitations of a reasoned and mechanical perspective.

Other forms of symmetricized meaning are the authoritarian regime and the fundamentalist discourse. In both, many meanings are fused together in what can be highly emotional and repressive responses and expressions. Symmetricization is thus the basis of primitive "common sense," or "folk" and traditional lifestyles. These unreflected lifestyles present singularities of habitual meaning and behavior that are fertile grounds for political or commercial propaganda and authoritarianism. Traditional and folk lifestyles are open to increased symmetricization because their structural basis is symmetricization. Using abstract moral imperatives such as "blood" and "soil" (employed in Germany in the 1930s), traditional lifestyles can be infused with cultural repressive meanings and behavior. Symmetricized meaning is therefore both susceptible to propaganda and is a powerful tool when used as propaganda.[7] Symmetricized propositions are evident in serious stereotyping in chauvinistic and jingoistic expressions, paranoid fears, and overdetermined generalizations, such as those found in religious, political, and cultural fundamentalism.

Bonding and connection are fundamentally important for family and community cohesion, but when these symmetrical relations are intensified into compound identifications of symmetricization, family and social formations can become difficult arenas in which overprotection, overcontrol, or emotional and rhetorical control is exercised. Symmetricization also gives rise to such ideal notions as the perfect body, perfect love, the sovereign nation, an individual or national identity, and a pure or elect culture. Sentimentality is therefore a typical symmetricized response, and romance and patriotic stories and songs common mediums for its expression.

Symmetricization can operates within the relationship between nature and culture, in various forms of "naturalism." A naturalist view fuses various features—observer, observed, interpretive genre, and specific cultural and historic conventions—into a singularity of empiricism in science, naturalism in art, or hegemony in a society. A symmetricized worldview is represented by the belief in, and reverence for, an impersonal, mysterious "fate."

Reason

A paradigm of reason pays homage to the second relationship of meaning, that is, by valuing nonsymmetry or difference as primary. I have called this framework of consciousness a paradigm of reason because the historical assumptions associated with reason appear to fit this orientation. A perspective of reason usually operates through binary reasoning, sometimes to establish an "objective" view in the pursuit of empirical methods. At other times a rationalist paradigm will produce a solipsistic argument, and at other times it is used in the pursuit of formal rules and principles that are valued above the unpredictable generic abilities of our own subjectivity. The reasoning inherent to reason is linear and usually involves the logic of dualism.

The following features are central to all perspectives of reason: differences are seen as exclusive; the explicit is always implied; so too is the visible. The perspective of reason often deals with the separate and dissociated, fundamental indivisible ontological substance (variations on the God particle). At other times reason assumes a reductive essence (essentialism) or a formal transcendental essence (difference in linguistics or Platonic mathematics). Nonsymmetry is the central value of the empiricism that deals with the explicitly visual as well as the reductive substance (in biology, the gene). Such substances are measured by explicit and differential calculus. The explicit, visually differentiated entity, such as a planet moving through a static but differential grid of space and time, is not only the mechanical image of the world, it is also a rational one. A perspective of reason therefore employs cause-and-effect models that traditionally are applied universally to social and personal actions, to biology as well as to the mechanics of physical entities.

The orientation of reason is reflectively differential, which means that difference is the automatic recursive gesture and the central value of this view of the world. The exclusiveness of this orientation comes from a tendency to erase relations of symmetry that in contemporary philosophical discussions are said to provide the concepts of structure, system, closure, similarity, and connections. More traditional forms of reason have attempted to erase or marginalize the symmetries of intuition and feelings by calling them irrational. The irrational is seen as an "antireason" that represents half of the binary pair rational/irrational. Such a perspective tends to objectify entities and formalize texts as differential elements belonging exclusively to (depending upon a materialist or idealist bias) either the physical world or to a formal transcendent zone.

Consciousness and unconsciousness are normally erased from an analytical perspective, but when consciousness is directly addressed, as with Karl Popper's "three worlds," it can be seen to operate as one explicit and differential element among other differential elements. The unconscious is usually seen as a problematic form of the explicit. For many rationalists the unrepressed unconscious, as unconsciousness, therefore does not exist. The same comments apply to meaning, which is normally erased or ignored in linguistic and semiotic analysis but which when addressed (Saussure and Derrida) will represent a differential value in a differential system of values. Traditionally the observer and observed are viewed as separate and different; likewise theory and practice, and learning and action. In the same way cognition, language, and the physical world are seen as separate from one another.

We find the valuing of difference in the traditional "criterion of rationalism." This criteria entails the static and foundational, substance concept of identity (Being), which in modern philosophy can be either absent or present. Difference is the prime value in the logic of noncontradiction (another feature of this criteria); it is said a thing cannot be both itself and not itself at the same time. Finally, the valuation of difference is also evident in negation, which is just another term for difference.

A rationalist methodology will always be rigorous in its commitment to finding more and more differences and finer and finer distinctions. A common rationalist attitude will be to treat every implied or explicit reference to symmetry skeptically. For reason this means that unconsciousness is usually erased, while symmetricizations are normally termed "irrational" and enthusiastically rejected; nonmaterial formative agents are said to be the stuff of superstition. Thus God cannot exist for reason.

The general privileging of difference has been a feature in traditional, mechanical, Newtonian and Euclidean scientific and mathematical practices as well as in deconstructive theory. Derrida has been more rigorous than most rationalists, for he has made it his mission to seek out and deconstruct the symmetries of traditional rationalism that come in various forms of closure, as in the case of "presence," "structure," and "system."[8] One effect of this kind of reasoned orientation consistently applied has been to establish an antirealist and also an antihumanist position.

This paradigm in relation to social behavior engenders a set of responses ranging from skepticism, alienation, and prejudice toward the "other" to liberal beliefs that stress individual autonomy and self-interest. This range is diverse, but it has a common dualist basis established in the certainty of self and other. This perspective represents the self as an ego that, in its striving for autonomy, is concerned with its own interests and differences from others. The dualism of self and other, assumed by the ego, is held in place by a set of mechanisms that defend the ego against the other. The ego defenses, when strong enough, can fracture the integration of the system in general of consciousness into seemingly unrelated islands of subjectivity that are then reflected in the alienating social relations of the individual.

This perspective represents the established paradigm in Western academic thought, both scientific and nonscientific. Such a paradigm of consciousness, though always manifestly disparate when applied, will nevertheless aim for the common "rigorous analysis" in line with the "analytical tradition." When pursued and consistently applied, reason produces the fragmentation of knowledge, a situation common to most Western universities, which preserve the identity of separate and limited disciplines.

The valuing of difference results in an aesthetic appreciation of surfaces and can also determine the organization of our perception of any scene, discourse, or style of life. For example, commercial toys like clothes, houses, furniture, and cars can fill up our lives to the degree they are centrally important, that is, they can become valued as objects-in-themselves. This form of consciousness will tend to be motivated by surface appearances and by social differences. Why is this? An exclusive focus on objects automatically emphasizes their differential, exclusive, closed, and superficial character. As objects are always different one from another, like linguistic units, an exclusive focus on them will tend to produce an exclusive and differential aesthetic.

If our analytical and philosophical tradition has been such that we view difference as having the only certainty, the connection and integration of anything will be problematic.[9] Such a scepticism has been refined in deconstructive writing to the degree that difference represents the "origin" of all meaning and language. From an isomorphic perspective, if the "grammar" of difference is pushed far enough (as it is in deconstruction), it can result in a schizophrenic kind of superrationalism.

The analytical perspective of reason is always narrow and decontextual. This perspective tends to objectify and "instrumentalize" meaning, so that the many features of the material analyzed are with difficulty connected to one another. Such is the case whenever we cite "objective facts," or believe that formal, surface appearances are divorced from a particular background context. Instrumental learning theory is a case in point. Instrumental learning means decontextualized learning, with a focus on practicing "the basics" and knowing the "correct answers," without regard for the generic abilities of students or the social contexts in which learning occurs.

The narrowness of reason is also shown in its compulsive adherence to formal rules, which often center around such cultural constructs as objectivity, facts, data, the empirical, and, in more fundamentalist forms, the word. We find this compulsion to rules in behaviorist and empirical psychology, in religious fundamentalism and formalist grammar, and in the elitism of all top-down bureaucratic approaches to organizational and corporate life. But such compulsion is also part of the general Western intellectual tradition that I am sure narrows every academic discipline, so that much effort is spent fighting border disputes on questions of legitimacy, field, and identity.

Obsessiveness with formalistic rules may be a Western intellectual predisposition, but the effects of this compulsion also appear to be reflected in some forms of neurosis and psychosis. I imagine that much would be gained if a study were made of neurosis and psychosis in terms of its rationalist characteristics, that is, a study of the insanity of reason. For example, anorexia nervosa contains many rationalist characteristics: an obsessiveness with rules, a compulsive desire for certainty through disconnection, the accumulation of data and facts, and an inability to deal with personal connections with family and society. Schizophrenia is another illness that exhibits, in an extreme form, many of the characteristics of reason—a hyperreason in which the closures of a "normal" personality are fractured by the neurological equivalent of a hyper-nonsymmetry.[10]

No doubt there are other forms of insanity that are cultural in character or in which the paradigm of reason plays a reinforcing role. One can imagine that the insanity suffered by some abstract expressionist painters in this century was due, in part, to their adherence to a rationalizing modernism. But these speculations draw me from my path, which is to examine the structural and nonsymmetrical aspect of reason.

Reason, therefore, is a narrowing paradigm that tends to erase linkages; when these *are* acknowledged, they will be framed as tentative, suggestive, and inconclusive. The so-called mind/body problem is a classic example of a difficulty produced by the viewpoint of reason and the orientation of difference. The unity of "mind/body" can never be solved by working from the perception of reason, for, the orientation of reason will always produce differences and fragmentation even where, from an isomorphic perspective, the evidence of integration is overwhelming.[11]

In structuralist and poststructuralist theory, the problematic status of difference has become a central question mark. Difference will remain problematic ("the possibility of conceptualization") as long as we view language and meaning from a paradigm of reason. Because reason tends to orient us automatically to difference, when the philosophical material we view is itself the notion of "difference," then we tend to end up with a philosophy of superrationalism, in which differences perpetuate themselves into the vertigo of Derridean *differance*. In some respects this deconstructive and centripetal tendency could be described as an overdetermined conscious awareness in the Freudian sense of having several causes that reinforce each other. As I have suggested, this centripetal process is a kind of insanity; not the insanity of irrationalism but one arising from the large-scale fragmentation of a hyperrationalism.

The worldview of the paradigm of reason is a mechanical and reductive one, where deductive and differential computations can manipulate indivisible and substantial entities. Cause and effect are unquestioned, and we stand mute before the Jung's concept of acausal relations and synchronicity. This is the world of the skeptic who believes passionately in material causes but never in the nonmaterial formative causes of cosmic consciousness.

The Isomorphic Paradigm

Two centuries ago the German poet, playwright, and philosopher Johann Wolfgang von Goethe proclaimed a new way of scientific thinking. It was to concentrate on the whole rather than on the part and on the form rather than on the elements of material substance. This holistic method of thinking went against the analytical traditional of the time and so found little favor in the prevailing mechanistic age of science.

I have called this last paradigm of consciousness isomorphic rather than holistic because I want to focus on the basics of relationships. It is relationships that construct and structure holistic thinking and perspective, and therefore it is relationship that should be emphasized through the concept of isomorphism. Because of the historical lack of definition of holistic thinking, this category of perspective is still in the process of definition; to a large extent this book represents an attempt to define it in relation to meaning, consciousness, and discourse.

The orientation of an isomorphic paradigm is less fixed than the two preceding perspective frameworks. The focus here is on the recursive integration of differences, which is a dynamic involving recursive symmetries. Simplified, this may be called the orientation of recursive symmetry. Recursive symmetry can be seen as emergent, transformational, formative, and generative depending upon the field to which it is applied. In terms of meaning, recursive symmetry is the basic recurring process by which all systems emerge, change, and evolve. From this perspective, everything can be a system, and every system is constructed from the relations of symmetry, nonsymmetry, and asymmetry.

The organizational order of this paradigm is captured in the term asynordinate, used in this book to refer to the two aspects of every system. These are the hidden foundations of every system that give rise to its explicit and surface features. These two aspects are always integrated, so that the explicit is always a feature of the implicit, and the visible is always a constituent of its hidden context. In the same way, the effable is always given meaning by the ineffable. We see an asynordinate character in the relationship between the quantum particle and its wave front, or at a macro level between the wave and the ocean, or the connection between the individual and the community, the object and event, order and disorder, the explicit and the implicit, and so on.

This asynordinate organization gives rise to determined but unpredictable results, or to orderly disorder. This kind of organization has relevance for the study of social problems that always have the character of orderly disorder and where social cohesion is possible only through the valuing of cultural diversity. This isomorphic approach therefore has relevance to public policies on ethnic diversity, immigration, cross-cultural communication, education, ecology, and the environment.

The asynordinate organization of an isomorphic paradigm also has a circular character. An isomorphic paradigm lays an emphasis on the deep, implicate context: the unconscious, hidden meanings and the prereflective feelings, sensations, and intuitions; and then on the surface appearances and constituent features that are integrated into their contexts. Conscious awareness is therefore located as an emergent, secondary semantic structure of subjectivity in the way that physical entities have emerged from the symmetry interactions of cosmic consciousness. In terms of the recursive integration of this paradigm, any observations should automatically include the predispositions of the perceiving subjectivity. This means that the generic and the specific, the eternal and the relative, are both active and constituent features of this perspective

The personal predispositions that an isomorphic perspective implies are those associated with patience, tranquillity, humility, and a reverence toward the formative, interconnecting web of one's life. Social interaction is therefore valued more highly than the reasoned concepts of individual and social identity. As the whole of life is a continuous field of active meaning and consciousness, any meaning generated by social interaction will affect the surrounding social field. As the relations generated by individual behavior or discourse are seen to have vibrational energy, they will generate what Rupert Sheldrake has called "morphic resonance." Morphic resonance is the vibrational effect of past patterns of activity on subsequent activity, involving "a kind of action at a distance in both space and time."[12] In Hinduism these nonlocal, symmetry effects are called Karma. Attention to consequences and the effects of past behavior is therefore an essential feature of this perspective.

The worldview produced by an isomorphic paradigm is recursive, contextual, and holistic. It sees an undivided universe, where everything is connected to everything else through the divine, formative relations of symmetry. This worldview rejects the materialism of reason and the fatalism of symmetrization in favor of an optimism that comes from empathy and compassion. The optimism is seen as the realization of our integration into a larger whole, as an awareness of the symmetry relations that connect us to others, to the environment, and to cosmic consciousness or God. The subjectivity of an individual is a reflection of the infinite symmetries of cosmic consciousness. We can, therefore, find God in ourselves, as well as God in every aspect of the universe.

These are summary, general descriptions of the three paradigms of consciousness, descriptions that could be expanded to fill a book. It should be stressed that while these paradigms are organizationally distinct and therefore have their own "logic," they are not structurally separate one from the other, for each incorporates the same three relationships of subjectivity.

General Features

Several general points can be made about these three paradigms of consciousness. The first is that a paradigm of consciousness does not imply a direct link to intelligence. In other words, a perspective paradigm is simply the organization of meaning, and these may be subtle or crude, gross or sophisticated. Taking on a paradigm of consciousness does not automatically ensure a high or low intelligence. Intelligence, as measured by intelligence tests, may be biased toward the paradigm of reason, thus reflecting the history of the discipline of psychology. This is of no great significance, either for the distinction between paradigms or for the development of isomorphic capacities.

The second point is that the three paradigms have a self-reflective asynordinate structure. It will be recalled that the overall basic structure of the system in general of subjectivity is an asynordinate system, having an unconscious (implicate)/conscious (explicate) structure. This double system loops back on itself recursively for each of the three organizations of subjectivity. For example, a symmetricized paradigm that has its orientation in symmetry will tend to produce reactive behavior and expressions that are unconscious of its own unconsciousness. This naïveté can be gross, as in fundamentalism, or sweet, as in children.

A paradigm of reason also has an unconsciousness of its own unconscious, but this is reduced in a series of logical assumptions about self-evidence by a process of rationalization, which takes as its beginning a dualism that challenges the asynordinate structure of subjectivity itself. This happens because an orientation of nonsymmetry creates an apparent division between conscious awareness and its own symmetricized unconscious. The distinctions made in conscious awareness are thereby privileged over the "unreason" of its own unconscious foundation, which is then often objectified as existing "out there" in others as some kind of irrationalism. Thus, reason will always have a symmetricized foundation that is rejected, ignored, or erased in the same manner that the logical always emerges from the illogical.

The asynordinate basis of an isomorphic paradigm of subjectivity breaks with the unconscious traditions of symmetricization and reason and produces a self-reflection of both its own unconscious, symmetrical foundations into which it integrates the explicit differences of reason. Together these combined meanings produce the systematic integration of differences within larger wholes or contexts. The interactions within and between the two aspects of the asynordinate structure of an isomorphic paradigm is fluid and flexible, and individuals who display these abilities can be said to be "in touch" with their own and their culture's unconscious.

The asynordinate structure of each of these three perspectives means, for example, that a symmetricized perspective will contain a repressed double dose of symmetry while having little ability to draw distinctions. In contrast, reason will be too able to draw distinctions and yet unwilling to accept its own unconscious and at times symmetricized foundations. Finally, an isomorphic paradigm has the capacity to accept its own unconscious foundations and the ability to draw distinctions, and integrates them both into systematic, holistic contexts.

These summary comments should be sufficient for most readers to realize that almost everybody will, every day, make use of all three paradigms of consciousness in relation to a range of actions, people, subjects, and events with which they come into contact. It is possible to have all three perspectives functioning in relation to a set of discourses or a difficult human relationship. For example, we can be prejudiced against our neighbor, colleague, or boss because they "appear" like the authoritarian figure who terrorized us as a child and then rationalize this symmetricization into a set of exclusions and justifications—"they are to blame for . . ."—and then finally discover through some self-reflection in prayer or therapy that the associations of the past can be integrated into less stressful present behavior.

However, in the main, a perspective paradigm refers to the predominant patterns of identification (symmetrical patterns) that an individual displays in any particular situation. The predominant patterns of identification in a symmetricized perspective are those associated with fear, pain, trauma, or unfulfilled desire, relief, and pleasure. At times these patterns can so overwhelm the individual that survival itself often appears as a daily struggle. Superstition is a word that can describe these kinds of identification patterns.

The predominant patterns of identification for reason are those associated with the construction of a personal ego. The ego is both a negative and positive construction. Its negative side comes from a variety of defense mechanisms that were identified by Freud and that are in the repressed unconscious, that is, they derive from symmetrization. Its positive side emerges from materialist identification patterns (which are also symmetricized meaning). These are patterns of identification that identify the individual with a range of material and even formal objects. Most common are the beliefs that "I am my body" and "I am my house, car, spouse and clothes." Also common is the belief that "I am my discourse." Bodily appearances and also the ownership of material "toys" like cars and houses are therefore of paramount importance for maintaining the ego. Freud's reality principle, which he said was the basic principle to the ego, represents a reasoned justification (a rhetorical strategy) for the perpetuation of these materialist and formalist identification patterns and so ties together the culture of reason with the psychology of ego. Materialist identification patterns have over the past three hundred years developed the clinical, specialist technologies and skills in a variety of different disciplines and fields. These specialized skills are appropriate for developing the skeptical approach to the world and generally for the working out of mechanical and engineering puzzles. Materialism is therefore a term that covers the identification patterns of reason.

The predominant identification patterns in an isomorphic perspective are those associated with reflection and self-reflection concerning the very links and connections that embed us within society, nature, and the cosmos. These are nonmaterialist identification patterns in which the individual self represents a nonmaterialist, formative, and therefore divine symmetrical unity constructed from past and present relations and interactions. The self is thus constructed from the same immanent, formative, and divine relations as cosmic consciousness. From an isomorphic point of view the path that identifies self with God or cosmic consciousness represents the realization of a basic reality, which is that the underlying reality of the universe is the formative panentheism of symmetry. Realization of this structure is therefore a consciousness that understands the nonduality of self and God. Illumination of self is therefore to realize God.

These nonmaterialist identification patterns can be developed into isomorphic abilities. Isomorphic abilities are those that make us able to recognize, understand, and construct connections, linkages, and similarities and to discern the generic in subjectivity and the eternal in cosmic consciousness. The generic features of subjectivity are those associated with the symmetrical structure of subjectivity and present in every social interaction. Such generic features become discernable in the "heart," in the feelings of love, compassion, and empathy and in the social interactive behavior of communication. A "heart" process that can increase the isomorphic feelings of love, empathy, and compassion is meditation that concentrates awareness on the symmetry of a divine form such as Jesus, Buddha, Muhammad, Ramakrishna, or Ammarchi.[13] Isomorphic abilities can also be developed through discursive practices that provide a focus for relations as well as through a range of spiritual practices.

Nonmateralist identification patterns are therefore spiritual patterns. Spirituality is a term that can describe the patterns of identification common to an isomorphic perspective.

While these three paradigms of consciousness usually refer to our predominant patterns of identification, it is also the case that these three perspectives can operate selectively in regard to such universal behavior as love. Love involves a recursive accent on the symmetry that binds two people, two organisms, or two particles. Thus a symmetricized human relationship will be one where there is a high degree of sexual and/or romantic infatuation. This kind of binding could perhaps be described as an unconsciousness of an unconsciousness of an unconsciousness. A loving relationship based on reason is one where the differences in a symmetricized relationship have become of paramount importance, so much so that infatuation has passed into open conflict. The differences that a reasoned perspective constructs are often so valued that they become the grounds for ending the relationship.

In contrast, an isomorphic loving relationship is one of empathy and compassion in which the differences of the relationship are accepted and integrated into the subjective patterns of the participants. Sexual partners who are young usually have a symmetricized or reasoned relationships, while parent-child relationships more often operate on some variation on an isomorphic relationship. These are oversimplified and traditional roles that are not always played out as suggested here because there is a wide range of other symmetricized patterns from the participant's repressed unconscious that can negatively and sometimes positively affect their bonding relationships.

The interconnections of these personal examples can also be extended to scientific endeavor. In other words, while an isomorphic perspective can be applied to some area of study, like the calculations of chaos theory, that does not mean this perspective will translate into the administrative practices of the laboratory or into the personal habits of the scientists, both of which may be rational and patriarchal (that is, symmetricized and rationalist). Thus the shifting from one paradigm to another in relation to different subject matters can occur as quickly as a change in context or situation.

Yet another way to look at these paradigms of consciousness is to use a mathematical analogy. For example, the three paradigms of consciousness are roughly analogous to three kinds of iterations we find in the various "feedback" mathematics of ecological and mechanical systems. To iterate a mathematical function is to use the output of one calculation as the input of the next in a feedback loop that is repeated endlessly. This is the kind of mathematics suited for dynamic systems such as population changes. An example from James Gleick's popular *Chaos* is the typical high-school calculation in which we suppose that next year's population of gypsy moths will depend entirely on this year's population; that is, thirty-one thousand gypsy moths this year means thirty-five thousand the next. The relationship between all the numbers can be captured in mathematical function: the population next year (x_1) is a function (F) of the population this year (x): $x_1 = Fx$.[14] Thus the whole history of a population becomes available through this process of functional iteration.

Functional iterations can map not only ecological movements but any dynamic system, yet different systems can behave differently. In one place, feedback can produce stability, as a thermostat does in regulating the temperature of a house, while in another, feedback can get out of hand. This happens, for instance, when a loudspeaker feeds back through a microphone and is amplified to an unbearable noise.[15] On the other hand, as Gleick tells us, "if the sound is small enough, it will die away to nothing."[16] It is possible to model these simple processes (*[a]* the amplification of noise and *[b]* the reduction of noise), one-dimensional with numbers by taking a starting number, multiplying it by itself, multiplying the result by itself, and so on. We will "discover that large numbers lead quickly to infinity: 10, 100, 10,000 . . . But small numbers lead to zero: 1/2, 1/4, 1/16 . . . "[17] The first series of numbers can represent amplification and the second series of numbers reduction of noise.

The iteration of systems with two dimensions becomes more complex, and, depending on the initial starting values and the rate of change, for instance in a population, the results can show either a move to extinction, a steady state, or finally a chaotic one. A chaotic system is one that shows complex behavior that is regularly irregular, or practicably unpredictable. Thus the effects of feedback (iteration) will depend on a range of factors related to the complexity of the system being measured.

These examples of mathematical maps have some rough similarity to the three paradigms of consciousness. For example, iteration in which the numbers run off to infinity can be seen as similar to the runaway differences of reason when assiduously and consistently applied. This is especially the case in deconstruction philosophy, in which there is an infinity of difference; a continual difference of difference of differences. This straight-line movement to infinity is also characteristic of the breaking up and fragmenting of knowledge into smaller and smaller bits, which is typical of the reductive and discipline-bound approach to knowledge. In a similar but reverse vein, symmetricization can be seen as the kind of meaning that fuses things together so that any set of numbers quickly reduce to extinction: zero. Both these kinds of iterations are one-dimensional and are therefore only partly analogous to the mechanical and reductive character of a symmetricized or reasoned consciousness.

Finally, the orderly/disorderly behavior of complex systems can be likened to the balance of an isomorphic consciousness where there is a continual flow of both order and disorder, of integrated differences in determined yet unpredictable outcomes. From an isomorphic perspective, randomness, ambiguity, unpredictability, and uncertainty are inherent and necessary aspects of the dynamics of consciousness. Yet so too are the traditional concepts of order and certainty. With an isomorphic perspective the traditional binary opposition between order and disorder no longer operates; instead there exists a synthesis and both-ness of order and disorder. The concept of an orderly disorder is one that squares well with the concept of the asynordinate. If something is asynordinate, there is an integration of unconscious and conscious meaning, of feelings with intellect and of differences embedded within symmetries.

From a rationalist perspective these are always exclusive pairs. From an isomorphic perspective all pairs, including such notables as body and mind, explicit and implicit, form and field, behavior and meaning, visible and invisible, ineffable and the effable, are linked and connected into an integrated both-ness, a togetherness. A both-ness is also apparent when referring to the relationship between consciousness and the physical world. The "in here" and the "out there" of a rationalistic view are connected together simultaneously as a unity of subjectivity. In the process of "seeing" the environment we make no distinction between "in here" and "out there." The symmetrical connections within the field of perception contain nonlocal connections or at least connections that have no specific location. A rationalistic view dismisses this continuity of perception and in its place constructs the naive realist distinctions of "in here" and "out there."

The ethics of an isomorphic paradigm are ones associated with what could be described as "the middle path." This is the path that tries to contain the disorder that comes from either too much symmetry (unconsciousness) or too much nonsymmetry (from recursive differences) while producing a balance of orderly disorder that comes from the recursive integration of differences. The effect of the integration of differences will be, in many cases, to downgrade the value of significant rationalist differences. Such is the case with the "in here," "out there" distinction. The value of this distinction is subsumed under the influence of the symmetrical connections that are dominant in the concept of a subjectivity that embraces everything. Within this unbounded field of subjectivity the distinctions of time, space, and object can be and are made, but they are made within the field itself, not as separate from it. In general the ethics of reason are based upon an either/or imperative, and such an essential tends to decontextualize all relationships by giving undue weight to differences and distinction. This is what happens with the exclusive concepts of "in here" and "out there"; right and wrong; fore and against.

Structurally, an isomorphic paradigm produces a balance of symmetry and nonsymmetry, a balance of unconsciousness and conscious awareness—in contrast to symmetricization, which produces too much unconsciousness, and reason, where there is too much striving for the explicit of conscious awareness. Both these extremes reconstruct either an oversimple or a mechanistic order that does not take the other into account. We find this problem both with the unconsciousness of too strong emotional feelings and expressions (too much identification) as well as with the cold detachment of conscious and mechanical reason (too much difference).

A public-policy example of the desire for too much symmetry (symmetricization) is the immigration policy of assimilation. An immigration policy of cultural assimilation is one designed to ignore or erase the cultural differences of the immigrant groups while advocating an idealized notion of mainstream cultural sameness or purity. In similar vein, a reasoned public policy will advocate cultural and ethnic separation. Such a policy can induce its extremes in the form of racism. The meaning of racism contains the overvaluing of simple and unimportant differences, usually the surface differences like skin color, dress, or etiquette. Neither public symmetricization (assimilation) nor social rationalism (apartheid) or their various combinations are adequate perspectives for social formations or public policies.

The only paradigm of consciousness that can accommodate the orderly disorder of any society is an isomorphic framework that allows for pluralism, the policy that provides national and social unity from cultural diversity. From this perspective framework no cultural group will have unity to the degree of an undifferentiated identity. No matter the ethnic background or its history, every group, whether small or large, local or national, contains the cultural differences of age, gender, family and personal history, religious interpretation, region, and many other important features. Diversity is therefore an essential feature of any specific subgroup, as well as of the larger community. The stability of this kind of community can develop from public policies directed at increasing tolerance of, and respect for, cultural diversity. These are the policies of pluralism, but they also represent the working out of an isomorphic paradigm.

These examples indicate that paradigms of consciousness relate to all and any phenomena. There is essentially no difference between the field of quantum mechanics and a social formation. Each area will be organized in some way by a combination of the organizing principles of symmetrization, reason, or isomorphism.

Self-Similarity

Ignoring for the moment the rhetorical power of discursive habits, traditional grammar, and the conventions of application, each of which work to preserve the traditions of reason, paradigms of consciousness can nevertheless be varied with a change of focus. A change in focus will in turn change the ordering and organization of behavior and information. Yet what is it that we can focus on to change our perception of the world?

One way to answer this question is with the term *self-similarity*. Self-similarity is often used in the literature of chaos theory to denote the concept of various levels of recursive symmetry. Self-similarity is a concept that couples together the various scales or levels of any dynamic system into an integrated whole. Self-similarity has sometimes been referred to in the past by Western philosophers and poets. Leibniz conceived of this kind of integration of various scales when he imagined that a drop of water contained a whole teeming universe that contained, in turn, water drops and further new universes.[18] Perhaps because of its natural poetic dimensions, self-similarity, operating between scales, has inspired some scientists to describe it in verse. The chaos that water turbulence shows is described here by Lewis Richardson in an intentional parody of Jonathan Swift:

> Big whorls have little whorls
> Which feed on their velocity,
> And little whorls have lesser whorls,
> And so on to viscosity.[19]

As this piece of doggerel implies, chaos theorists view dynamic systems as having a coupling effect that integrates perturbations on the smallest scale with recursive self-similar patterns at a larger or macro level. Hayles gives another example of this linking between the smallness and largeness of contexts in the cracking of a whip. "Imagine a bullwhip moving at just the right frequency so that a small twitch of the handle is transmitted into larger and larger waves all the way to the end, causing the whip to emit a loud, satisfying crack."[20]

But self-similarity is not only a concept used in chaos theory and in the mathematics of such symmetry practices as "renormalization." The meaning of self-similarity is also relevant to the isomorphic feeling of empathy. Empathy is a complex interpersonal response in which the differences of the relationship are accepted and integrated into the subjective patterns of the participants. To develop empathy it is therefore necessary to practice self-similarity—the practice of seeing oneself in others and others in oneself. Empathy is impossible without the practice of self-similarity. In addition, an empathetic response tends to reproduce empathetic responses in others in a self-similar way. The production of empathy is perhaps similar to the way a small twitch on a bullwhip handle is transmitted into larger and larger waves. In other words, empathy is perhaps the key to a chaos theory approach to social interaction.

Chaos theory examples of self-similarity relate to physical systems. This exclusive use of self-similarity has tended to lead scientists to view it as an exclusive property of physical system but not as a feature of subjectivity or social interaction generally. Thus also the scaling factor identified as belonging to complex physical systems has not been seen to include the recursive dynamics of meaning and consciousness. It is because of this general erasure of subjectivity from their calculations that we can say that chaos theorists are not working entirely within an isomorphic paradigm but often import the deleterious effects of reason into their work.

It is possible, however, to demonstrate the integration of subjectivity in relation to any so-called static system. A static system can become complex by the simple expedient of foregrounding subjectivity rather than system type. In other words, the dynamic structure of the act of observation—involving the system in general—will become more apparent if the structure of the perceiving consciousness is taken into account. To illustrate this point we use the following example from Hayles's article "Chaos as Orderly Disorder." Hayles tells us to

> imagine two paintings, each showing an open door through which is revealed another and another . . . One way to think about the doors in these two paintings is to focus on the particulars of the repeated forms. Suppose the doors on the first painting are ornately carved rectangles,

> whereas the second painting shows doors that are unadorned arches. If we attend only to these shapes, the paintings may seem very different. But suppose we focus on the recursive repetition of the doors in each painting and discover that in both paintings, the doors become smaller at a constant rate. Through this shift in focus, we have found a way of looking at the paintings that reveals their recursive similarity.[21]

This example illustrates how a simple change of focus, of orientation, can provide us with a comparison of two perspectives and two interpretive methods that apply to the same set of static images. Hayles advises us to "focus on the recursive repetition" of the forms within the pictures, and it is this simple advice that alerts us to the possibility of comparative paradigms of consciousness and to the consequent involvement of subjectivity itself. In terms of subjective experience these changes in orientation come about from a focus on the differences of a system or a focus on the recursive similarities; the self-similarities of a system.

It should be noted that in the question of "which focus" we do have a choice and therefore do not have to accept Hayles's advice, for we could focus instead on the differential particulars of the forms, and in so doing we would assume the established orientation of a rationalist perspective. In other words, we are free to choose a perspective paradigm (once we know about perspective paradigms) and thereby construct the world in a particular way. But we are not absolutely free to construct the world as we would wish, for we cannot change the invariance of the underlying structure of symmetry, nonsymmetry, and asymmetry of the system in general. In other words, we are not free to step outside the subjective field of meaning. Rather, we are only free to choose one of three paradigms of consciousness.

In this example of Hayles's it is apparent that the nature of the system perceived will help or retard an isomorphic perspective. So-called static systems tend to retard and dynamic systems to assist an isomorphic perspective. But this is not the only factor; another is the notion of paradigms of application. Our training, our language, our traditions of analysis, our methods of experimentation, and our theories have a great influence in maintaining the perspective of reason. But in spite of the rigidity of these cultural influences, if the physical system itself is complex enough or the method we use recursively symmetrical enough so as to force us into a focus on self-similarity (as chaos theory tends to do), then these influences may combine to disrupt the perspective or reason and also indicate the need for an isomorphic paradigm.

It is difficult for those of us who write about an isomorphic perspective to totally throw off the old expectations of reason. It is therefore possible for complex systems to be interpreted both isomorphically and from a point of view of reason. That is, specific areas of the complexity may be seen from a reasoned perspective while the main body of the system is viewed isomorphically. This is a common occurrence with much of the literature of new physics, chaos theory, and quantum mechanics, where the involvement of a perceiving consciousness is usually marginalized or not acknowledged. Instead, recourse is usually an implied rationalist consciousness.

When this happens, the work suffers to the degree to which a paradigm of reason is used to investigate the complexity of dynamic systems. For example, a paradigm of reason based on an assumed predictable order cannot account for the chaotic recursive symmetries of complex systems, nor can it account for the involvement of a perceiving subjectivity that functions as part of the system under investigation. This cross-paradigmatic approach is common to chaos theorists and even parts of the work of David Bohm, who has done more than most to apply isomorphic principles to science.

Conclusion

In this chapter some of the sight lines for different perspective were examined. These perspective we called paradigms of consciousness, as distinct from Kuhn's notion of paradigms, which assume a set of practices and applications. Kuhn's paradigms were therefore called paradigms of application.

Three and only three paradigms of consciousness were proposed. These are symmetricization, reason, and isomorphism. A paradigm of consciousness implies that organization is distinct from structure. Thus while the structure of subjectivity remains invariant, being constructed of the relations of symmetry, nonsymmetry, and asymmetry, the organization of subjectivity can take three possible directions. These directions are determined by the orientation of the perspective and the orientation in turn is determined by the central recursive value of the perspective. Thus symmetricization has the pattern of undifferentiated recursive symmetry that leads to compound identifications; reason has a recursive pattern of nonsymmetry that leads to dissociation; while an isomorphic paradigm has recursive patterns of asymmetry that leads to the integration of distinctions. Much more could be written about these three generic perspectives and their relationship to the relativity of cultural paradigms of application. However, this discussion will have to wait for another time.

Chapter 8

Awareness and Creativity

In this chapter I want to explore the semantic basis of conscious awareness and reflective self-consciousness and then look at some of the implications of this structural approach to creativity.

Contrasting Views

Traditionally conscious awareness is viewed as a featureless unity of attention that is the essence of a Cartesian cogito. Sometimes this featureless cogito is associated with the notion of intentionality or will and freedom of choice. Quite separately, rationalists have viewed their own philosophical position as one concerned with the self-evidence of reason. It is the isomorphic view here that these are not two separate areas covered by the disciplines of philosophy and psychology. A Cartesian cogito and the self-evidence of reason refer to the same subject matter: the ingredients of conscious awareness.

The "self-evidence" of reason has tended to blind its adherents to anything that lies beyond this so-called self-evidence. In the same way rational psychologists have been unable to name the features of attention and conscious awareness, for to do so would have meant standing outside the self-evident boundaries of their philosophy—in other words, being located in an isomorphic epistemology. From an isomorphic perspective self-evidence or reason refers both to *(a)* the value placed on the distinctions, differences and nonsymmetries of a rationalist philosophy and *(b)* the distinctions, differences, and nonsymmetries of conscious awareness. These two sets of differences are identical.

The failure of reason to articulate the ingredients of self-evidence or to name the features of attention and conscious awareness has occurred because of a blindness in this perspective. This blindness involves the failure to contextualize nonsymmetry as an emergent relationship that in turn has stemmed from the more general inability to account for the meaning of symmetry. Unable to account for both of these, the perspective of reason has fallen back on their symmetricization and called the resulting fusion of meaning "self-evidence."

Reason's self-evidence thus refers, not to the certainty of a coherent logic, but to the self-evidence that we have conscious awareness to socially interact. Now distinctions and differences may also be self-evident as a general class in awareness, but specific distinctions are usually the sticking points of all disagreements. The problem with self-evidence is that it hardly ever is.

But describing self-evidence in these terms is to indicate also the semantic features of conscious awareness. These are the relationships of nonsymmetry, which are usually called distinctions and differences. Reason's hegemony, which identifies an idealized internal logic with the differential processes of conscious awareness, has simply prevented researchers from adequately accounting for these processes. We see evidence of this in both Freud's and Matte-Blanco's work.

Matte-Blanco did not use or conceive of nonsymmetry ether as distinction or difference in any structural and psychological sense, and I see no Freudian psychological system that represents these relations. In contrast, nonsymmetry is a term that has great importance for semiotics and discourse analysis, for its meaning involves the important function of discrimination, of differentiating. It will be recalled from our discussion of Benveniste's approach to meaning that the notion of discrimination represented half of the functions involved with meaning, the other half being integration. Difference is vital and a fundamental term for structuralism and poststructuralism and for linguistic philosophy generally; so given the present semantic approach it is no surprise that nonsymmetry is seen as an integral aspect of any description of consciousness.

The Structure of Conscious Awareness

From a structural point of view the relation of nonsymmetry represents the negation of symmetry, in other words, the negation of unconsciousness. The negation of unconsciousness is awareness or attention. Awareness is produced and functions as a system of marks and distinctions—nonsymmetries—which simply *are* the characteristics of conscious awareness. When referring to conscious awareness it is therefore not necessary to, in addition, refer to objects "in" awareness. The binary concept supplied by "object in awareness" implies that awareness is distinct from the objects it constructs, in the manner that a light, which illuminates objects, is separate from them. This is a mechanical image and totally inappropriate. So also is the common image of awareness as a kind of "mental screen" on which outside objects are projected.

In contrast, to suggest that awareness is created by the construction of nonsymmetrical relations indicates that awareness is created by and through the very process of constructing differential relations. The patterns and objects we create "in" awareness are therefore created from differential relations. Awareness and objects "in" awareness are therefore identical conditions. In this isomorphic sense we are the objects of our perception. Relations of nonsymmetry are therefore the constituent feature of patterns and objects within awareness as well as the constituent features of conscious awareness. Because of this identity of structure (we are our perceptions), the constituent features of conscious awareness will always be visible, explicit, and differential relations (as Freud has said of the perceptual or W system). In contrast, the constituent features of unconsciousness will always be hidden, implicit, and symmetrical relations.

Some appreciation of a rationalist approach to conscious awareness can be derived from John Frisby's description of how the visual system is involved in "seeing" an object.

> A rectangle is built up of various features, usually four corners and connecting edges or lines. Perhaps when we "see" a rectangle, what has happened is that the visual system has identified certain corner/edge/line features as being present and then discovered that these are so arranged with respect to one another that they satisfy the scene description "rectangle present." If the features had been arranged differently, then a different scene description would have been arrived at (such as "square present" for example).[1]

This rather mechanical description has several shortcomings. First, the description implies that a system (a rectangle) is built from several *different* features. This is a model of dissociation in which the primary conditions of the seeing are separate and differential. Such a description is common to reason, which always places the value of difference first in any order of preference. With such a set of preferences Frisby is then unable to account structurally for the unity of the rectangle, so he resorts to the time-honored image of the homunculus, the little preexisting awareness within the visual system that is able to "then discover" that the pieces fit together into a rectangle. Like most rationalists who use dissociation models to discuss integrated processes, Frisby has to call on some form of already existing, undefined awareness to mysteriously perform this integrating function. A conscious deciding agency at this level of prereflective consciousness is structurally impossible and is generally considered, even by those who use it, to be a fallacy.

From an isomorphic perspective it is not the different features of a rectangle that are primary but the unconsciousness of symmetry transformations. These are a mimetic transformation that deals with the whole system of the rectangle, not its individual features. Individual features, corner/edge/line, are secondary characteristics of the rectangle, features that have been extracted out of the system in a reflective process involving representation. Like all rationalists, Frisby has reversed the order of abstraction to present an erroneous, differential features analysis that cannot explain seeing a rectangle or the conditions of awareness.

Neuropsychology has explained the limitations of a features analysis not in terms of the semantic relationships of conscious awareness but in terms of experimental results related to pattern recognition. For example, Karl Pribram concludes that a features analysis has three kinds of problems: (1) features are distinctive, (2) the richness of perception cannot be accounted for by features, and (3) behavioral discrimination does not correlate with information from feature detectors studied by microelectrodes.[2] These limitations led Pribram to look for other explanations of pattern recognition and image production in perception. By emphasizing the neural characteristics of a wave front that operates at synaptic junctions and the interference patterns that affect them he formed the view that here was a holographic pattern. Thus, by analogy, Pribram hypothesized that perception could work like a hologram.

This was a radical shift from feature analysis and one that is in line with the present isomorphic approach to consciousness. The difficulty I have with the proposal that perception works like a hologram is that a hologram is badly defined in terms of the relationships that construct it. As mentioned in chapter 3, holograms are formed by splitting the light from a laser into two and then bringing the two beams together by shining them on an object. The interference patterns from the two sources of light produces the three-dimensional image on the photographic plate. The image is called a "hologram" because every part of the image reconstitutes the whole image.

Now while this is a fascinating set of processes, it does not provide us with a language, a set of interrelated terms that are useful in explaining the creation of images in perception or, for that matter, describing conscious awareness. Terms like "wave fronts" and "interference" patterns do not describe or strongly imply the relations that must constitute them. On the other hand, an isomorphic approach that relies directly on the terminology of relations has the potential to draw out much more information about holograms than previously anticipated and also to be in a better position to come to conclusions about pattern recognition and conscious awareness. For these reason I have made little use of holographic analogies and images in this study.

It should be remembered that the recognition of any pattern is a complex set of processes. These processes do not represent a mosaic of points that could make up a photograph-like image on the cortex of the brain. Such a view results from a framework of reason. Rather, pattern recognition represents a set of transformations that are generated from the hidden symmetries of the perceptual processes and from the symmetries from the environment as well as from the symmetries of memory. There are therefore at least three major systems interacting in the recognition of any pattern. But in addition patterns themselves suggest organization, organization within and between these three systems. This is a great deal of organization and, at a physical level, a great deal of brain stimulation.

How is it possible for that amount of coordinated organization to occur and not malfunction? This is a question of mechanics. But of course the relations that make up this high degree of organization are not mechanical but formative, constitutive, and structural. In other words, pattern recognition itself is not a mechanical affair but represents a formative and structural aspect of a larger subjectivity. Pattern recognition therefore has some of the symmetry potentials of cosmic consciousness. What are these potentials?

Pattern recognition tends to be a problem phrase because while it says nothing about conscious awareness, conscious awareness is always implied as the state in which pattern recognition occurs. If this is the case, then what part does conscious awareness play in pattern recognition? This is too difficult to answer, so we need to rephrase the question in terms of relations: "What part does nonsymmetry play in pattern recognition?" A clear answer is now possible because we know that conscious awareness is constituted by relations of nonsymmetry.

In terms of relations, pattern recognition involves constructing relations of nonsymmetry and combining them in various ways to form patterns. The part of this complex process from which the relations of nonsymmetry emerge is the hidden, implicate area of the process. The part of the process in which the relations of nonsymmetry are constructed represents the explicate area of conscious awareness. The part of the process that combines these differences into patterns is again the hidden, implicate area of the process. There are no boundaries between the parts, as each flows into the other in seamless transformations.

Pattern recognition is therefore an asynordinate set of processes, and it is misleading to view it as simply a mechanical process that happens within the state of conscious awareness. Pattern recognition creates conscious awareness in the process of creating patterns. Without pattern recognition there is no conscious awareness because there are no relations of nonsymmetry. We are therefore, so some extent, the objects and patterns we construct in perception.

In terms of the structure of subjectivity, conscious awareness does not and cannot stand alone as pure intentionality, rational mind, or a Cartesian cogito divorced from its hidden foundations of symmetry. The nonsymmetrical marks that create conscious awareness are always a secondary process constructed out of "the sea of symmetry," the unconscious, or what Freud called "the general basis of psychical life." In this sense conscious awareness systematically emerges from unconsciousness through the semantic processes of the negation of symmetry, a "non" symmetry. Conscious awareness is thus not a solo individualistic state of free will that can feed off itself; rather it represents the secondary and transformed set of meanings within a three-valued recursive structure of asynordinate subjectivity.

Asymmetry

As we stated in chapter 3 in reference to cosmic consciousness, the third tier in this story of emergent transformations is the relationships of asymmetry. This is the case both for cosmic consciousness and for subjectivity. This means that the structure of subjectivity is recursively isomorphic with the structure of cosmic consciousness.

As suggested above, conscious awareness is part of an asynordinate system that integrates differences into patterns. The category of a system is one that represents a series of asymmetrical relationships, or in other words, a system is a series of integrated differences. This use of the term *system* as asymmetrical relations cuts across the assumed division between implicit and explicit meaning or unconsciousness and conscious awareness. Asymmetrical relations include both these categories in integrated ways that make it impossible to separate them. Thus to identify conscious awareness with asymmetrical relations, in the manner that Matte-Blanco employs, is to import reason into the picture and make divisions where there are only transformations.

What we need to describe subjectivity is not a series of exclusive terms that overlay divisions in an otherwise integrated system but a phraseology that foregrounds the integrations and transformations of the overall system in general. The term *system* has the semantic structure of recursive asymmetry; it is therefore structured not just by a single difference or a single irreversibility but by a recursive set of asymmetrical relations. This complex process is therefore not linear or singularly differential of the "one thing at a time" variety but is more like a fractal process where the same ingredients occur again and again, from the smallest enclosures of meaning to the overall universal context of cosmic consciousness.

John Eccles has correctly said, "There is a *unitary character* about the experiences of the self-conscious mind."[3] The unitary character of which he speaks does not come from some single, featureless intentionality or transcendental cogito but represents the very symmetrical character of the recursive system itself; of the interlocking relations within systems within systems. Self-consciousness is therefore asynordinately structured by recursive asymmetrical systems. "Self-consciousness" is perhaps a better term to use to represent this integrated system than *conscious awareness* or *attention* or just *awareness,* because each of these latter terms implies some kind of separate state of subjectivity.

It should be obvious from this discussion that the picture of subjectivity being presented here is not the Freudian topographical one, in which we pass from the state of "ordinary thinking" into the basement of the unconscious. Nor is subjectivity presented as an image of a ladder that we can climb to gain higher and higher levels of awareness. Nor is it represented by an image of a container that is empty at birth and can have varying levels of fullness in adult life. Subjectivity here is not represented as some kind of quality that is produced by 10 percent of the brain and that could be expanded by another 90 percent, nor for that matter is consciousness represented as a free agent having deconstructive dimensions; nor is it seen as featureless or dismissed as "autoeffect."

Subjectivity is here represented as a series of recursive enclosed systems: a fractal-like series of systems symmetrically coupled together. The degree and complexity of coupling will relate to the degree of reflexivity and, finally, to the reflexivity, that is, the self-consciousness of the overall system. The image of consciousness presented here is of an overall circular system of repeated and continuing systems coupled together and made manifest through a myriad different types of closures. The system in general of subjectivity therefore has the structure of a system, and the smallest recursive feature of consciousness is a system, and such a system represents a "package" of meaning and maybe is a functional object like a linguistic unit: the morpheme or the phoneme or a simple figure like a line, square, or circle. But these smallest systems have the same structure as the overall system in general of subjectivity. Their differences are a matter of scale, not structure.

This recursive overall system does seem to have several levels that are distinct and that would be produced by different types and degrees of closure. For example, we could call the complex automatic processes associated with sensory stimulation the first and prereflective level of subjectivity. The second level of subjectivity relates to the act of pattern recognition and involves prereflective consciousness as well as the construction of conscious awareness. A third level of subjectivity involves formal symbolization, naming, and categorizing and the reflective actions of communication. At this level subjectivity is involved in the processes of conscious reflection.

A fourth level of subjectivity develops out of the other three by the isomorphic coupling of system coupled to system coupled to system until the overall system of meaning and subjectivity is included. At this final and fourth level we have self-reflective consciousness where the meaning of self is realized as an aspect of the larger context of cosmic consciousness. Thus for consciousness to be self-reflective there has to be a level of recursive complexity that includes the originating system from which subjectivity has emerged. Self-reflective consciousness is therefore a further isomorphic development on the preceding three levels. This view of recursive sets of interlocking systems is very different from the static and often exclusive image of the "mind's eye." The mind's eye is here represented not as a featureless cogito, but as an asynordinate system of implicit and explicit reflexive relations that operate within the whole fractal-like system of consciousness that includes subjectivity.

From an isomorphic perspective a subjectivity that is self-reflective is valued more highly than a subjectivity that is not. The reason for this valuation is that the recursive logic of an isomorphic perspective privileges recursive symmetry; it therefore follows that self-reflection is privileged over linear, reductive, and mechanical reasoning. The path of self-reflection is a path of recursive symmetry that leads back to the originating cosmic form of symmetry. Thus in a contradictory sense that is typical of isomorphic systems, a consciousness returned to the value of its own unconscious foundations of symmetry will become, not less conscious, but self-reflective. Self-reflective consciousness is therefore an isomorphic perspective that is in touch with its own recursive structure, for it takes into account, in some final sense of reassessment, the system's own practices in the present and in the past, the residue of which has often accumulated as a repressed personal history.

Self-reflective consciousness is therefore a perspective of greater recursive symmetry and of higher complexity than the perspective of reason, where reflection does not include so-called subjective and spiritual elements. Self-reflective consciousness increases the level and amount of recursive asymmetry, for it involves a series of primary, secondary, and tertiary interlocking systems within a system that loops back on itself holistically. In an autopoietic sense, self-reflective consciousness is a self-creating system because it is self-reflective. (The biological term *coupling* can be used isomorphically with recursive symmetry, for both terms imply a continuation without separation between one series of asymmetrical structures and the next.)

A subjectivity that is not self-reflective, evident in most adults much of the time, is a consciousness where the coupling of system to system stops at some point that is not the system's own self-reflective processes. This kind of subjectivity will usually have the differential reflexivity of reason and an orientation that erases its own symmetrical foundations. Thus even though this subjectivity of reason has an asymmetrical structure, self-reflection will not be a feature of this perspective, and the organization will not take into account its own structure. This perspective of reason, produced through learning or habit, emanates from a focus on surfaces and differences and a fixed identification of self as material existence. Reason's orientation leads to splits and separations of subjectivity and a view of the world that is similar. Reason's symbolizations always reinforce the values of this perspective and also tend to prevent the recursive complexity needed for self-reflection. Under these conditions consciousness is evident, but not self-reflective consciousness.

Self-reflective consciousness operates to the degree to which the overall system of subjectivity contextualized the others three levels of subjectivity into coherent images of self. This means that while we may be self-reflective in a range of areas of our lives, such as work and leisure, we may not be so dispassionate when it comes to our spouse, children, religion, or politics. In the first case we maybe self-reflective to the degree of knowing our weaknesses and strengths at work or in sport, and being comfortable with this level of professionalism and detachment. In the case of our spouse, children, religion, and politics there may be little self-reflection of these relationships that create resentment, anxiety, fear, and paranoia. We may therefore employ an isomorphic perspective for only a small amount of time because at other times we are using a perspective of reason or symmetricization to respond reactively to the world.

If this is the case, then self-images will tend to be ego constructions. Egos are created from symmetricized defense mechanisms that are reinforced by patterns of identifications with material existence. Ego responses are therefore reactive and exclusive and non-self-reflective and tend to lead to disagreement and breakdown in personal and social relationships. Ego responses do not integrate past experiences with present actions but try and separate the two by erasing and repressing the past. In contrast, self-images that emerge from an isomorphic perspective are positive and not reactive for they are based on a set of symmetrically recursive and integrated relationships where the organization of subjectivity coincides with the structure of subjectivity. This coincidence represents an "enlightening" process that goes beyond individual subjectivity to take in the recursive relationship of cosmic consciousness and subjectivity. Isomorphic self-images are therefore essentially spiritual.

Symmetry Transformations

This brings us to the role of symmetry transformations. The whole *system in general* of subjectivity operates by way of symmetry transformation. Each of the four levels of subjectivity is made distinct by the symmetry transformations that produce them. Symmetry transformations will be different from each other depending upon their place within the whole system in general of subjectivity.

A symmetry transformation is a concept familiar to science and mathematics. The mathematician Joe Rosen refers to symmetry transformations in terms of any change that can be brought to a system. Even simple changes involve symmetry transformations; for example, a square is transformed by rotating it on an axis perpendicular to the plane of the square.[4] In science, symmetry transformations are as diverse and imaginative as the possible systems upon which they might act. For example, geometric transformations can be those changes in shape, size, displacement, rotation, and reflection of a single form such as a square. A system might be symmetrical in respect to the interchange of its parts. A symmetry transformation of time might involve the displacement or reversal of the order of events or a change in terms of the time interval or periodicity.

From the present point of view a symmetry transformation is not an independent property of objects in either the physical world or some formal abstract world of Platonic mathematics (as Rosen and many other scientists assume). A symmetry transformation is an emergent function of subjectivity that constructs the systems we take to be the independent systems of the physical world. These so-called physical (and mathematical) systems are not independent but are symmetrical with the transformations by which subjectivity operates. A symmetrical transformation of the kind Rosen speaks about is thus a triple symmetry transformation, involving a type 1/2 system plus a symbolic and conceptual type 3 system. It will be recalled from chapter 4 that a type 1/2 system represents the environment, or, in terms of the present discussion, the second level of subjectivity concerned with pattern recognition. A type 3 symbolic and conceptual system is the third level of subjectivity.

A symmetry transformation is a concept similar to Noam Chomsky's notion of a grammatical transformation. If we view a sentence as a system, then a grammatical transformation is a symmetry transformation. In relation to a sentence, Chomsky comments that "the deep structure is related to the surface structure by certain mental operations—in modern terminology, by grammatical transformation."[5] This is simply a description of a symmetry transformation in relation to a sentence. Both grammatical and symmetry transformations imply a complex system of hidden relations from which emerges a surface structure.

A symmetry transformation is different from a symmetrical relationship and a symmetricized relationship. Symmetrical relationships are the foundation of all meaning and consciousness as well as material existence, and there is always the formative and immanent potential for transformation to occur at any time. In a symmetricized relation there will have been in the past the formation of a group of symmetry transformations that now have been temporarily hidden by an imposed and repressing overlay of symmetry. This configuration (or syndrome) represents the repressed unconscious.

In contrast, a symmetrical transformation has the following features: *(a)* (deep) unconscious, hidden relations of symmetry from which emerge *(b)* a differential surface structure. In other words, a symmetry transformation involves the transformation from symmetry to nonsymmetry and then the transformation to asymmetry. In the strict sense of these basic relations, a symmetry transformation always involves a double transformation that moves from symmetry to asymmetry.

Symmetry transformations are ubiquitous, for they operate both in a hierarchical dimension, that is, as a change from one level to another, and "horizontally" within the space and time of any particular level. For example, symmetry transformations are closely linked to the mathematical notion of "scaling." The concept of scaling represents a hierarchical linking of different levels of systems within an overall system: a series of levels within levels within levels. Scaling can relate to particular fields of study; for example, in contemporary evolutionary biology, genes, individual organisms, species, and families of species represent different levels or scales in the overall perspective for evolutionary biologists. Each of these systems (genes, organisms, species, and so on) represents a different scale that is connected to the other larger or smaller scales in the overall system.

Scaling gives emphasis to the integration of systems of different size, even in economics, where, for example, the small-scale day-to-day fluctuations of cotton prices can relate to larger-scale monthly changes. This analysis was carried out by Benoit Mandelbrot,[6] who, as a chaos theorist, was interested in the kinds of regularity of change that occur across scales. This regularity (called self-similarity) across different scales represents the scaling of recursive symmetries, a scaling where the symmetries of one system at a particular level will be isomorphic with the symmetries at another level. I propose that the kind of hierarchy of structure inferred by the concept of scaling occurs in the field of meaning that links cosmic consciousness to subjectivity and discourse. In this sense, discourse and subjectivity represent a set of scaling processes that reflect similar processes at work in the universe.

A symmetry transformation is an asynordinate operation in that it is a process that brings new meanings into awareness (brings to light). New meanings develop out of hidden meanings, and so new meanings make us conscious of something of which we had previously been unconscious. For example, bringing some new meaning into awareness will necessarily involve a change to the already operating unconscious patterns from prereflective behavior, memory, or the repressed unconscious. In other words, a difference (a nonsymmetry) has to be brought about from within the already operating unconscious complexity of symmetries. This series of processes can be described as a "symmetry transformation" and from a psychological point of view involves a "becoming conscious" of new and different meanings. In this sense the processes of learning and "becoming aware" of something always entails symmetry transformations.

Creativity

The psychological process of "becoming conscious" of previously hidden meanings is a learning and an enlightening process, for what is brought to "light" (conscious awareness) is some aspect of a system (perceived or conceived) that was hidden within the symmetry of the systems structure. When we see more of the picture or a larger picture, we are involved in the process of learning. A symmetry transformation is therefore an isomorphic function that can either open up the closure of a system to display its internal constituent parts or transform a constituent into a context having other constituents. This is a dialectical and synthesizing set of processes that value interconnectedness and the notions of plasticity and flexibility of consciousness. Transformational learning is therefore learning that has outcomes concerned with flexibility and processes focused on symmetry transformations.

This learning process is akin to those often thought to occur in imagination or creative thought. Traditional rationalist philosophers often have difficulty with imagination. They have sometimes seen it as providing the connection and the synthesis for mental representations that are believed to be atomistic and individual. Seen from this perspective, imagination provides the unifying and connecting structure of consciousness. Yet imagination has also been understood as the reflexive capacity for creativity and the production of novelty in art, literature, and science.

Both these approaches to imagination—as the unification of consciousness and as the production of reflected new meanings—are part of the rationalist, philosophical tradition left by Kant that still affect attitudes today. For example, Mark Johnson believes that an adequate account of meaning awaits a comprehensive theory of imagination.[7] For Johnson and philosophers like him, neither has yet happened. Yet from an isomorphic perspective, the reverse is the case: an adequate account of imagination can occur only within a comprehensive theory of meaning.

The two ways in which imagination has traditionally been viewed has some relevance to the general functioning of symmetry transformations in subjectivity. For example, in a unified, isomorphic structure like subjectivity, different kinds of operations, such as prereflective sensations and emotions, perceptions, and reflective and self-reflective concepts and symbols are not separate, individualistic occurrences. Rather, they represent various aspects of an overall integrated and asynordinate system. This being the case, subjectivity will be structurally integrated at every level from prereflective to self-reflective consciousness. That integration is brought about through the operations of symmetry transformations. Symmetry transformations, therefore, not only take place at the edge of conscious awareness, as a creative process of discovery, but the unity of consciousness itself is constructed (unconsciously) by transformations that are symmetrical. They emerge naturally from the inherent symmetrical structure of subjectivity, which is a reflection of the symmetrical potentials of the cosmos.

Subjectivity, therefore, has its own symmetrical principles of unity and organization within itself. This unity is not so much a process of the imagination as a structural unity that emerges from, and is transformed by, the foundation principle of symmetry. In this sense the general concept of a symmetry transformation provides the link between the traditionally perceived dual role of imagination, that is, unification of consciousness, and the production of new meaning or novelty. From a contemporary viewpoint, however, the term *imagination* does not seem to square well with notions of an inherent structural unity. It fits more closely with traditional views of creativity that relate to the development of new and novel meanings.

However, I see creativity and imagination as overlapping terms that refer to the same set of symmetry transformational processes. As symmetry transformations occur throughout subjectivity, they will have slightly different effect in different areas. For example, if we take the structure of the isomorphic sign (see chap. 5) as representing the several "vertical" levels of subjectivity, then each of the three levels emerges out of the context of the preceding level through symmetry transformations. Thus the symmetrical foundations of subjectivity, prereflective subjectivity, which we represent as the first level of the sign, has emerged, by a long and complicated evolutionary history of symmetry transformations, from the structure of cosmic consciousness. The second level of differential percepts (in linguistics they are called signifiers and in psychology pattern recognition) emerge out of the unconscious, prereflective, symmetry-transformational processes of sensation and memory. The third levels of selecting and combining signifiers (transforming them) result in the emergence of images and concepts (signifieds).

Now these "vertical" levels also operate "horizontally" through a time sequence, and so symmetry transformations can occur between levels but also within levels. For example, they take place within the third level of the sign, with concepts, "in our imagination." A symmetry transformation of concepts is thus an imaginative activity that does not directly engage with the body's present prereflective motor and sensory transformations that usually form a feedback loop to the concept being transformed. This "vertical" disengagement from the specificity of present sensory and motor stimulation (transformations) enables the general, the generic, the infinite, and the eternal to be conceptualized.

The creativity needed for problem solving is slightly different from the predominantly horizontal transformations of an imaginative activity. Problem solving needs both the engagement of vertical transformations that locate an individual within a specific set of circumstances and situations as well as horizontal, conceptual transformations that provide generality to the specific. The balanced, dialectical interrelationship of these two, the general and the specific, are the essential ingredients from any successful problem solving. If one line of symmetry transformations predominates over the other, too much empirical information swamps the principles that tie them together; if overgeneralizations drown the specificity of local conditions, then difficulties and problems will not be successfully resolved.

Imagination and creativity are therefore similar processes and both involve symmetry transformations. Both entail the construction of new differences within the context of any given system. A system can predominantly be associated with the "horizontal" dimension of concepts, "in imagination." Alternatively a complex system can be creatively embraced through horizontal transformations of generality interacting with "vertical" transformations of specificity.

As imagination and creativity represent different forms of symmetry transformations and as symmetry transformations are structured by relations and relations are always formative, constitutive, immanent relations of structured meaning, then imagination and creativity will also be concerned with potentials and possibilities. Such potentials and possibilities relate to the formative, constitutive, and immanent potentials of symmetry. For creativity the potentials of symmetry represent the meaning potentials of subjectivity that are hidden from conscious awareness but that can be transformed by symmetry transformations. Creativity therefore involves a feedback process of intuition, in contact with its own unconscious foundations, to the degree that these potentials of unconsciousness can be called up into conscious awareness. Intuition in this sense means a reaching into the felt meanings and sensibilities of the implicate order of subjectivity and transforming them into explicit and differential communication.

Inhibiting Factors

If symmetry transformations are the structural basis of imagination and creativity and if consciousness operates by symmetry transformation principles, why is it that some people are more creative or imaginative than others? This question points to some of the factors that reduce and inhibit creativity. These are the same factors that prevent an isomorphic paradigm of consciousness from operating. This linking of paradigm with creativity implies that imagination and creativity are framed and supported by an isomorphic perspective—and also that creativity is enhanced and stimulated by the use of an isomorphic paradigm. Thus learning and creativity may both be assisted by methods and procedures that formally focus the student's attention on the symmetry transformations of subjectivity. This is the approach taken with transformational learning and generic ability-based learning.

The factors inhibiting creativity and symmetry transformations relate to ritual, taboo, tradition, convention, and political and ideological conservatism. These are the habits or symmetricization and reason, but before we discuss how these paradigms of consciousness inhibit creativity some comments need to be made about habit.

In recent years Rupert Sheldrake has upset biologists around the world with his metaphysical theory that the laws of nature are really habits of nature and these habits are built up and organized into fields by what he calls "morphic resonance." Morphic resonance is a collection of habits; a repetitive pattern of behavior that has a tendency for continuation under stable environmental circumstances. According to Sheldrake, habits built from morphic resonance are so strong that they are the reason why species develop as they do; why, for instance, offspring have similar forms to their parents. While I do not want to comment at any length on Sheldrake's thesis here, his approach does make use of isomorphic linkages and connections and therefore is sympathetic to the present approach. He makes out a strong case for the power and influence of resonance (symmetries) in developing habits that he sees as the main feature of invariance in evolution. Sheldrake considers that new habits come about by creative responses to the environment and that evolution as a whole involves an interplay between creativity and habit.[8]

Sheldrake's theory, however, suffers form the lack of a definition of creativity and a lack of structural content for his notions of habit and morphic fields. Given these two omissions, there is a tendency, in my opinion, to treat creativity too lightly. From an isomorphic perspective of consciousness, the resonance of a morphic field is the resonance of a field of the implicate order that is structured by the relations of symmetry. Morphic fields therefore will have the resonance, the vibrational energy of recursive symmetries. The resonance of recursive symmetries is a well-appreciated concept in music and need only be extended to all relations for us to be able to conclude that form-generating fields (morphic fields) have a resonance that has emanated from the recursive symmetries of their particular structure.

Such an extension is not unjustified. As we have seen in earlier discussions, all relations are relations of meaning and consciousness and thus have generic attributes. One of those attributes is clearly resonance. While this characteristic is highlighted in music, this is not to say that it does not exist in other mediums, like mathematics or communication. Music, mathematics, and communication are systems that generate nothing but relations and therefore nothing but meaning, resonance and consciousness. Resonance is a characteristic of each of these mediums, but because each is different from the other in a variety of ways so also is the meaning, consciousness and resonance generated by each. We may therefore extend Sheldrake's concept of a morphic field to include music, mathematics and communication, for each has a complexity of forms that resonate from their recursive symmetries. Morphic fields are thus not the exclusive province of biology.

Sheldrake proposed that the development of *new* morphic fields results from the "creativity" of Nature, that is, from innovations in organisms that occur in response to genetic or environmental changes. Creative responses to new environments are apparently a striking feature of plants and animals. Yet while Sheldrake has the necessary morphic explanation for the building up and development of new habits, he has no explanation for creativity. Such an explanation has to be structural and it has to fit within a larger context of the symmetrical structure of habit.

From an isomorphic perspective such an explanation is forthcoming. Thus while the meaning and energy of symmetry is the foundation of consciousness, both cosmic and in individual subjectivity, it is also the foundation of all biological life. But in addition, recursive symmetry can provide the structural content of habitual behavior and the reason why morphic fields are able to resonate. Following on from this, the transformation of symmetry into nonsymmetry provides the cosmic, transformational principle by which new habits begin. Thus differences of behavior once they have emerged from the symmetry of previous habit will tend to perpetuate themselves through the symmetry of repetition. In other words, they become new habits.

From an isomorphic perspective, recursive symmetry operates not only within subjectivity but also at all micro levels of cells, genes, atoms, and particles. Because each of these levels are interconnected to the others by symmetry transformations, along with having symmetry transformations operating within each level there will be a complexity of symmetry and nonsymmetry forces interacting to influence the outcome of any particular habit or potential new behavior. Creative behavior in the fullness of these forces is a very complex affair. The stability of habit will therefore always be in constant interaction with its possible transformation by creativity. As Sheldrake states, this is the case in nature, but it is also the case in subjectivity where the stability of symmetricized and reasoned habits interacts with the transformations of creativity.

This brings us back to how paradigms of consciousness relate to creativity. As suggested, difficulties arise in subjectivity when habits become fixed and unchanging. This happens in a variety of ways, for example, when the prereflective sensory habits of the body are imported wholesale into the conceptual level of subjectivity; in other words, when we symmetricize by identifying ourselves too closely with material objects. Habits also become cemented into place when we identify ourselves with symbols. In this way we develop an ego with its consequent defense mechanisms that seek to protect it from change and transformation. These kinds of fixed habits will respond sympathetically to repetitive social practices like rituals, conventions, and traditions that have themselves becomes rigid and unchanging. In these cases creativity is stifled because symmetry transformations are suppressed.

Symmetricized and rational habits in all their various forms are the strongest inhibiting influences to creative symmetrical transformations. For example, a symmetricized paradigm will not lead to a great range of transformations because of the rigidity and lack of symmetry transformations that emerge from this perception. Symmetricized systems will not so much be transformed as welded together into naturalized unities, or in more emotive ways, act as catalysts for displacing or condensing a range of other meanings. These fusions of meaning are not symmetry transformations but previous transformations that have now been suppressed by overgeneralization. For example, a symmetricized view of symbols can overgeneralize to the extent of bringing the transcendent and the instrumental together, as in the literal "word" of God that is an instrumental truth.

A greater range of symmetry transformations is possible through the use of a paradigm of reason. A rational subjectivity, however, works against the grain of symmetry transformations. From the perspective of reason it is the exclusive nature of differences that is significant rather than the integration inherent to the concept of transformation. Given that there is an infinite variety of forms in the universe, it is perhaps little wonder that the Western intellectual tradition of reason has tended to focus on the exclusion of these differences and has come to believe in the rigor of discrimination as an essential aspect of the rational mind. In the broad sense, an intellectual focus on material objects can lead to materialism in its many variations. In the case of rational idealism the focus has been on the logic of internal consistency and distinction as a self-evident value. But both idealism and materialism have the same exclusive valuation of difference, which is the orientation of the paradigm of reason. Thus Jacques Derrida, caught within the perceptual sight lines of reason, can confidently say that when we privilege consciousness over *differance* it "is the ether of metaphysics."[9] Such a viewpoint, which assiduously pursues differences, insofar as this pursuit takes place, is not self-conscious or self-consciously reflective or sympathetic to symmetry transformations. This is not to say that the history of reason has not produced significant symmetry transformations; it has. Rather, it is to point out that the differential "logic" of the paradigm of reason inherently works against the notion of symmetry and the production of creative symmetry transformations.

In contrast, an isomorphic perspective reinforces the "natural" symmetrical interconnection within and between different systems. An isomorphic perspective therefore places value on the recursive, unconscious, and implicit possibilities of a system and therefore the symmetry transformations that any system can generate. As no system exists on its own but represents one aspect of an integrated series of interlocking systems, the perspective inherent in the processes of symmetry transformation leads on to a self-reflected consciousness. An isomorphic perspective will therefore tend to automatically produce the creativity of symmetry transformations and a consciousness that is self-reflective.

Conclusion

This chapter discusses some of the common and major features of subjectivity by relating some cultural practices to these features. The self-evidence of reason is isomorphic with conscious awareness, with both being structurally linked through the relation of nonsymmetry. Reason places a priority on nonsymmetry, and conscious awareness is structured by nonsymmetry. This isomorphism has prevented reason from articulating its own so-called self-evident structure or that of conscious awareness.

Reflective consciousness has an asymmetrical structure, and self-reflective consciousness emerges from an isomorphic perspective that is a perspective in touch with its own unconscious and spiritual foundations. I discuss symmetry transformations in terms of the interactive dynamics of subjectivity and in terms of creativity and suggest that a subjectivity in touch with its own structural foundations of symmetry (an isomorphic perspective) would be more creative and imaginative than a subjectivity dominated by reason or symmetricization.

Chapter 9

The Unity of "Mind" and Body

In this chapter we develop an isomorphic interpretation of the unity of neurological activity and subjectivity. The traditional rationalist approach to the unity of "mind" and body is fraught with difficulties. This is because the paradigm of reason foregrounds differences and is predisposed toward finding them, yet unity is a concept involving reciprocal symmetries. A perspective predisposed toward finding differences will therefore have great difficulties in establishing unities. It is perhaps no wonder that philosophers like Karl Popper and John Eccles say that they doubt we will ever understand this problem.

> The problem of the relation between our bodies and our minds, and especially the link between brain structures and processes on the one hand and mental dispositions and events on the other is an exceedingly difficult one. Without pretending to be able to foresee future developments, both authors of this book think it improbable that the problem will ever be solved, in the sense that we shall really understand this relation.[1]

In contrast to this kind of pessimism, I maintain that an isomorphic paradigm, which predisposes us to recursive symmetry and therefore unity, can provide a positive answer to this relationship. The nature of this positive answer is that neurological activity in the form of excitation and inhibition are the actual semantic relationships of symmetry, nonsymmetry, and asymmetry and that through these the recursive systems in general of subjectivity are established. The unity of "mind" and body is therefore complete and absolute. But what is the path by which we arrived at this simple conclusion?

A Description of the Problem

Historically the philosophical problem of the unity of "mind" and body has suffered from the lack of a detailed description of the features of the "mind," in contrast to the great deal of information we know of neurological activity. Because the model of the "mind" has usually been of a Cartesian cogito, which is largely featureless, it has been difficult to establish isomorphic patterns between it and the brain. Psychology has been of little assistance to philosophy. Traditionally there are no cognitive units or "cells" that could be matched up to brain activity. Given these omissions, it is no wonder that the identity theory of mind and body has had difficulties.

In the mid-1950s U. T. Place advanced a rational hypothesis that consciousness was a process of the brain. To argue this proposition Place described two forms of identity that he called the "is of definition" and the "is of composition."[2] The "is of definition" refers to the identities produced by discourse, that is, created by the rhetoric of syntax (specifically the verb *to be*). The "is of composition" refers to the apparent isomorphic patterns that belong to phenomena. (These two categories clearly separate the observer from the observed and language from the observed.) It is the second kind of identity, the "is of composition," which Place saw as representing the identity of mind and body.[3]

Because the featureless state of the cogito creates difficulty in establishing relationships linking "mind" and body, Place used analogies to argue his case for the "is of composition." The basis of Place's argument is as follows: when we observe and describe such phenomena as a cloud, a flash of lightning, or even consciousness, our descriptions will be different when the phenomenon is observed from two different locations. Yet while the descriptions, observations, and locations are different, they are nevertheless descriptions and observations of the same event. The identity of the phenomenon is thus maintained through a composition of different observations and descriptions.

To illustrate the composite character of this kind of identity, Place reminded us that the phenomenon of a cloud is the same whether we observe it at a distance and call it a "cloud" or whether we are enveloped in it and call it "mist" or view it scientifically as a "mass of tiny particles in suspension."[4] Place argued that the three different descriptions from three different observation points do not subvert or disrupt the unity of the phenomenon observed nor the identity of description and phenomenon. Place also uses the example of lightning; lightning observed at a distance is ordinarily reported as a "flash of lightning," but when observed by someone who has learned about the motion of electrical charges through the atmosphere, the phenomenon can be reported as a "motion of electrical charges." The two reports are different and come from different contexts, but that does not make the reports false nor the identity of lightning different from the reports. Place would argue that here we have an identity of composition; an identity composed of a variety of different descriptions, but an identity of phenomenon nevertheless.

It is from this position of composition that Place launches his argument for the identity of mind and body: "We realize that there is nothing that the introspective subject says about his conscious experiences which is inconsistent with anything the physiologist might want to say about the brain processes."[5] In other words, the description of introspective experiences by an individual and the description of neurophysiological activity by physiologists represent different descriptions of the same phenomenon. Yet the question arises from this composition: are the introspective report and the physiologist's description referring to the same processes? While there is nothing inconsistent in these two descriptions, is there anything that ties them closer together so that their identity is assured?[6]

While Place states that there is nothing inconsistent with an introspective report of conscious experience and anything a physiologist might want to say about brain processes, the possible gulf between these two kinds of description could be very wide. This is so because there are many more possibilities for establishing differences that seem contrary than opportunities for establishing similarities that are isomorphic. In other words, while Place's basic hypothesis of an identity may survive, the lack of fine-grained detail of introspective reports, or a description of the structure of the cogito, makes the link between the "mind" and brain so disparate as to defy anything but the most tentative conclusion. Albert Shalom, a critic of Place's thesis, goes so far as to declare that the identity theory is simply motivated by the belief "that everything *should* be reducible to the world of physical reality."[7]

Popper concludes that a retreat from the identity theory has become fashionable, "a retreat into what may be called a 'promissory materialism.'"[8] According to Popper, promissory materialism offers better possibilities in the future when more is known about the functions of the brain. But in the meantime, "the identity theory is hardly compatible with ordinary language or with common sense."[9] Such comments are common in Popper's criticism of the identity theory.[10]

The gulf between introspective reports and the descriptions of known brain activity is also reflected in Eccles's account of neurological activity in *The Self and Its Brain*. This gulf leads Eccles to postulate a dualist-interactionist hypothesis that proposes that "the experiences of the self-conscious mind [have] a relationship with neural events in the liaison brain, there being a relationship of interaction giving a degree of correspondence, but not an identity."[11] Eccles characterizes the self-conscious mind as having a unitary character, concentrating now on this, now on that aspect of performance. This description is then compared with a general description of neurological activity: "The operations of the brain can be understood only in terms of *neuronal prodigality* in the constitution of myriads of spatio-temporal patterns."[12]

These two descriptions appear as totally incompatible, for how can a diversity of brain events give rise to the unified conscious experience of a global or gestalt character? Eccles answers the problem by adhering to his dualist-interactionist hypothesis; thus, it is not the brain that gives the self-conscious mind its unitary character but the mind itself. His ideal and rational approach is complete: "the experience of unity comes, not from a neurophysical synthesis, but from the proposed integrating character of the self-conscious mind."[13] The mind is therefore separate from the body, but it does have some kind of interaction with the body. Eccles seems to be content with this kind of comment, which leaves the mind as a mystery.

The gulf between this kind of description and the brain's propensity for diversity is large enough for the differences to appear totally irreconcilable. Given this level of incommensurability, it seems surprising that Eccles goes so far toward identity as to talk in terms of mind-brain interaction. This example, along with the others, indicates that if a theory of the identity of mind and brain is to move beyond the most general of terms proposed by Place, there has to be some detailed description of mind activity.

Directions for a Possible Synthesis

Like many other writers, both Popper and Eccles have suggested that we need to know more about the brain's activities. Contrary to this view, I would argue the reverse: that the multitude of books written on neurophysiology, which describe in subtle and complex detail the numerous micronetworks of the nervous system, attests to a level of information far superior to anything written about consciousness. The problem has not been with descriptions of brain activity, it has been with the lack of descriptions of consciousness and subjectivity. For Freud and almost every writer since him, conscious awareness is a featureless state. How is it possible then to demonstrate on the one hand the identity of a featureless state and on the other a system so full of features that it is near impossible to keep up with the reading?

Contrary to the introspective record, which is the established approach to "mind," we do not deal directly with introspection here. Rather, the proposal is to show how the set of cosmic relations of symmetry, nonsymmetry, and asymmetry that form the architecture of subjectivity are coextensive with neurological activity.

Conditioning

In proposing a connection between meaning's structure and neurological activity I have drawn upon the seminal work of the nineteenth-century Russian physiologists Ivan Mikhailovich Sechenov and Ivan Petrovich Pavlov. In an effort to describe the internal world of thoughts, feelings and desires by scientific and materialistic methods, Sechenov and later Pavlov settled on a model of the organism as an integral whole inseparable from its environment. For Pavlov, the organism, which included homo sapiens, was a dynamic system: "an enduring system, in mobile unstable equilibrium."[14] Even though he was a materialist, Pavlov had a partial holistic approach, for he saw that the internal and external environments of the organism represent an integral totality. This unity of "inner" and "outer" is brought about through the operations of the central nervous system that receives stimulation from both of these sources. In addition, Pavlov held that there is an integral unity of mental and somatic, subjective and objective. Throughout his life Pavlov was a persuasive exponent of what is called psychophysical monism.[15]

Central to both Sechenov's and Pavlov's views of the organism (a term that covers life from the sponge to humans) were the terms "excitation," "inhibition," and the "reflex." The first two of these terms are still used widely today by psychologists and physiologists to describe the functioning of cells in the nervous system of many organisms. The term *reflex,* George Millar suggests, came from Descartes, who referred to the automatic actions that the soul was unable to control voluntarily.[16] In the early nineteenth century medical examinations began to reveal the neural pathways of reflex actions. From this time on reflexes were considered to be unconscious, so that when Wilhelm Wundt started his new psychology in the closing decades of the last century he took it for granted that it would have nothing to do with reflexes. This distinction between unconscious reflexes and the conscious, voluntary mind persists even today in spite of Freud's showing that much of mental life is unconscious. For instance, while Popper sees a difficulty in linking our bodies to our minds, he does nevertheless feel inclined to regard the identification of unconscious mental states with brain states as "a very important conjecture."[17]

In the fields of psychology and physiology, reflexes have become associated with the processes of conditioning and the pursuits of American behaviorism, epitomized by John Watson, who believed that "no introspective inferences about any mysterious mind-stuff were required."[18] What began as a Russian distrust of subjective criteria for experimentation ended in the cul-de-sac of American behaviorism, which virtually erased the subjective. Such a journey from Russian physiology to American behaviorism has tended to categorize Pavlov pejoratively as a behaviorist. He was not a behaviorist, for he accepted the existence of inner mental experiences and devoted most of the last decade of his life largely to psychiatric problems.[19]

Pavlov is known, however, for his discovery in the early part of this century of the conditional reflex. Experimental work with dogs to investigate the way their digestive glands functioned led Pavlov to make the distinction between a conditional reflex and an unconditional reflex. The difference is observed between a dog's salivation when bread is put in its mouth (a natural or unconditional reflex of the digestive system) and salivation induced by the sight of bread (the conditional reflex). The conditional reflex is considered not as inborn, but as something the dog has learned to do. Pavlov's great discovery of the conditional reflex meant that he devoted the rest of his life to improving and refining this knowledge. When in the final years of his life he turned his attention to psychiatric problems, Pavlov realized that language introduced enormous problems into the explanation of conditioning. There was a need for new explanations, so he added a further distinction to unconditional and conditional reflexes. The conditional reflex in animals he viewed as the "first signal system," while humans had evolved another, a "second signal system" of verbal symbols.

In 1932 Pavlov presented a summary of his second signal system to the Academy of Sciences of the USSR.[20] He saw this system of conditioning underlying mental activity and the production of symbols as the basis of a materialist science of psychology. While both Russian and American physiologists conducted numerous experiments on the second signal system, the psychological results never lived up to the promise that Pavlov had held for a truly scientific psychology. In a communication read to the delegates of the XIX International Physiological Congress in Montreal in 1953, K. M. Bykov, one of Pavlov's coworkers, seemed to justify the lack of results, saying that "such a complicated field of investigation as the study of the higher nervous activity has, of course, not been completed. . . . It is not easy to create a principally and methodically new science concerned with the activity of an organ whose study of human thought started twenty-three centuries ago."[21]

From the present isomorphic perspective Pavlov's difficulty in producing worthwhile results on mental states derived from both his materialist approach and from the crudity of his materialistic terms. Terms such as "conditioning," "unconditional reflexes," and "conditional reflexes" do not well represent the features of introspective reports or the characteristics of the unconscious, the repressed, or subjectivity generally. These terms do, however, have a utility in relation to materialist traditions of physiology. The problem that Pavlov faced was of paradigm and terminology. Thus the second signal system, while attractive from an evolutionary point of view, does not suggest anything about meaning, consciousness, or unconsciousness, elements that are at the very core of mental and symbolic life.[22]

The same criticism can be made of Humberto Maturana's systems of autopoiesis. While both Maturana and Pavlov have very similar views of the "organism" (for Pavlov) and "autopoietic systems" (for Maturana), both these materialist scientists have had difficulty in describing in any detail the nature of the subjective. For example, Maturana proposes that all living systems, from sponges to humans, have a cognitive domain. Such an intriguing proposal, however, is not supported by a detailed description of a cognitive domain, defined as "the domain of all interactions, in which an autopoietic system can enter without loss of its identity, is its cognitive domain."[23] This may be agreeable to biologists when they confront a frog, but it does not help us with the difficulty of subjectivity or the unity of mind and body. The concept of a cognitive domain for all living systems automatically implies an identity of mind and body. But this inference goes no further than that; it infers an isomorphism but does not describe its character. And there is no description of its character because there is no autopoietic description of consciousness or meaning.

In conformity with traditional linguistic views, Maturana places meaning in what he calls the "linguistic domain." As such, he sees no connection of meaning with neurophysiology. "Time and meaning are effective factors in the linguistic domain, but as relational entities do not have neurological correlates in the operation of the nervous system."[24] I will now elaborate on this correlation between meaning and neurophysiology.

Relations from Neurons

The basic question that has been asked many times by those skeptical of identity theory is, "How do neurophysiological processes become mental processes?" This question needs to be rephrased isomorphically: "How are neurophysiological processes isomorphic with the relations of symmetry, nonsymmetry, and asymmetry?" To answer this we need to go back to the work of Sechenov.

Predating Pavlov and the language of conditioning, Sechenov in the 1860s saw the concepts of inhibition and excitation as central to developing a materialist theory of psychic activity.[25] These two terms, "excitation" and "inhibition," which have been used so widely and with such utility since, have fertile implications for meaning and consciousness. Without reiterating Sechenov's theory of how inhibition and excitation function physiologically to determine and produce psychic activity, I intend to use these two terms as they are used generally in physiology today. Yet I also want to foreground their semantic and psychological potentials, which Sechenov clearly saw in this theory.

This emphasis is achieved by asking, "What relationship accompanies each of these states of excitation and inhibition?" It is proposed here that when neurons are in a state of *excitation,* symmetrical relations are produced, and that when excitation is *inhibited,* nonsymmetrical relations are the outcome. This does not mean that excitation produces symmetrical relations or that inhibition produces nonsymmetrical relations. Rather, it means that the materialist terms *excitation* and *inhibition* are replaced, without loss of meaning, by the nonmaterialist terms *symmetry* and *nonsymmetry.*

The state of excitation is therefore also a state in which symmetrical relations are created, and the state of inhibition is also a state in which nonsymmetrical relations result. The isomorphism of these two sets of terms is absolute in the sense that they refer to the same processes and the same phenomena. Their difference are due to a history and perspective paradigm of different disciplines. It is not due to a difference in system type. Their unity is one of composition, in Place's sense of composition. The unity of these two sets of terms therefore heralds the unity of nonmaterialist subjectivity with the materialist body.

In general this means that the neurological activity that has traditionally been called excitation and inhibition provides the materalist conditions for the production and organization of the complex relationships necessary for subjectivity. It should be noted, however, that the proposal that neurons produce the relations of subjectivity needs to be qualified to the extent that there is no simple, mechanical cause and effect implied by this unity. This may seem to be the case, however, for if neurons are physical and if they produce relationships, does this not mean that neurological activity is the material basis of subjectivity? This may be the conclusion of some who hanker after a materalist view of mind, but it is reached only by ignoring the larger cosmic context in which subjectivity arises, as well as ignoring the reason neurons fire. These reasons are complex, and explanations of biological and electrical interactions can be carried on through a range of levels until we find ourselves back at the point where Heisenberg has argued that particles emerge from fundamental symmetries.

From an isomorphic perspective, the materiality of neurological activity is only the explicate feature of a larger system of asynordinate processes. Neurons have themselves emerged from the symmetrical interactions of the implicate order of nature. They are, therefore, secondary effects of a nonmaterial, implicate order: an order structured by symmetrical relations. Neurons play the role of providing the gross material, organizational conditions for the active agents (relations) of subjectivity to achieve locality, focus and systematic unity. Neurons do not by themselves bring into existence unique relations, nor do neurons produce an epiphenomenal by-product called relations.

Rather, the nervous system simply provides a physical vehicle for the interactions of subjectivity to operate during the life of the body. In other words, explicit material objects, like nervous systems, can and do produce relations, but only because cells have themselves been formed and come into existence through the interaction of fundamental symmetries. In addition, neurons produce relations only in conformity with the system of relations already existing in the cosmos. This conformity is not a matter of chance, but one of extension, for the relations of subjectivity are of the same order (asynordinate and formative) as the divine, animate relations of cosmic consciousness.

The primacy of relations is also evident from the circular effects relations have on neurons. Thus the meanings created through neurological activity can and do affect the same neurology. Even in small ways such as hand-to-eye coordination, the relations of subjectivity are affecting neurological activity. This nonmaterial "causation" is happening all the time, in motion as well as in reflective action—for example, when we order and organize our bodies in relation to environmental conditions or when we reflect on something and take action based on those reflections. The whole field of meaning, which includes matter, individual subjectivity, and cosmic consciousness, is circular, interconnected, and interactive. While the present chapter deals with a narrow focus of the relationship between neurological activity and subjectivity, this focus should not be seen in a decontextualized manner as if it represents a mechanical and materialist cause-and-effect model. The larger holistic context is absolutely essential for understanding the manner in which the relations of subjectivity are produced and organized by the physical activities of neurons.

Neuronal Relations

The case for the equivalence of meaning and neurological activity is supported by the inherent priority of excitation over inhibition. As is the situation with symmetry and consciousness, excitation is a primary activity of neurons, and inhibition is a secondary activity. It is secondary insofar as inhibition impinges on excitation through a series of already existing processes (depending on the type of inhibition) that "shunt off" the excitation.[26] In other words, we cannot have inhibition until there is excitation. There is, therefore, a hierarchy in relation to these two physiological mechanisms, a hierarchy that is identical to the one that integrates symmetry and nonsymmetry. As I have already argued, nonsymmetry (difference) is a secondary process dependent on the primary one of symmetry. There is, therefore, the same emergent relationship operating with physiological mechanisms as there is with the semantic relationships of consciousness.

A further factor is that when a series of neurons fire and display excitation, apart from all the physiochemical reactions produced there will also be produced, if the excitation is uninhibited, a series of symmetrical relationships. But how are symmetrical relations produced by neurons? It is suggested here that the relationship of symmetry is produced automatically and involuntarily, from the point of view of the structure of meaning, between any neuronal point of stimulation and the uninhibited points of its effect. This is not a simple linear relationship but can involve a single input and a multitude of symmetrical effects. If there is no inhibition in this relationship, there will be no modification or negation (it is an all-or-nothing effect) between the separate points of stimulation and effect. Therefore, these separate points will simply be symmetrical in relation to any introspective notions of time, space, speed, or any other differential factor.

From an "observer's" position, a neuronal excitation always travels a distance at a certain speed and takes a finite amount of time. But these conscious and differential factors are illegitimate in terms of the structure of meaning of the individual involved in the stimulation. These differential factors represent the abstract calculations performed in the consciousness, specifically the representational system of conscious awareness, of the observer. These differential factors associated with stimulation are nonexistent for the participant whose neurons are being stimulated. Yet the terms observer and participant imply two classes of people, which is incorrect, for the observer is also the participant. The difference of the two classes is the difference between the processes of abstract representation and the prereflective excitation of neurons which produce these representations.

This is not a problem of infinite regress but one of perspective paradigms. From the isomorphic perspective of this study, it is taken for granted that nothing exists outside of meaning, that distinctions about the "inside" and "outside" of meaning are false. The biologist or physiologist "observer" who calculates the differential factors of neuronal excitation is using irrelevant calculations for the microstructure of meaning that is occurring in his subjectivity. If these differential factors are used in relation to meaning, then the "observer" is projecting his or her own conscious awareness down into the prereflective body and imagining that there is a "little person," a homunculus, there at the level of neuronal excitation.

From the point of view of the structure of meaning, an uninhibited excitation produces a relationship between the beginning and the end of the excitation, and that relationship for the participating subjectivity is one of symmetry. It cannot be anything else.

Now physiologists commonly see neurons metaphorically. Neurons are often described as receiving, conducting, and transmitting information, information that "can be likened to the transmission of Morse code along a telegraph line."[27] From the present perspective of meaning, there can be no message that can be likened to Morse code received, integrated, and transmitted along a nerve, from collector to affecter surfaces. From the present position, the "information" (sometimes called a signal)[28] that an uninhibited nerve cell is supposed to transmit is no more than the "message" of symmetry, which is the same as saying that a relationship of symmetry has been produced between collector and affecter. The present view thus denies the metaphoric speculation that some kind of macro, reflective consciousness (a homunculus) is already operating at a cellular level.

The concept of information as a message or signal implies that a conscious awareness somehow constructs the representational content of the "messages" or "signals" that travel the transmission lines of the nervous system. For information to be information it has to have representational form, it has to be a symbol for something else. But at this microlevel, meanings are nonrepresentational; they are simply the micromeanings out of which representations come. In other words, at the level of neurons, signals, messages, and information do not exist beyond the meanings of symmetry, nonsymmetry, and asymmetry.

Taking into account the character of an isomorphic perspective, we can say that uninhibited excitation, which can involve a whole range of various kinds of neurons, produces symmetrical relations as complex as the structure of the neurons. The relationship of symmetry is therefore equivalent to the mechanism of excitation.

Inhibition and Nonsymmetry

The proposal that neurons give rise to the relations of symmetry when action potentials are uninhibitedly conducted along their lengths is a relatively straightforward hypothesized equivalence. The situation is not so straightforward when it comes to the production of nonsymmetry by the processes of inhibition. The term *inhibition* has been used in a variety of different ways by Pavlovian physiologists, and at least three types of inhibition are recognized by contemporary physiologists. These three are presynaptic inhibition, direct inhibition, and inhibition from interneurons.[29]

Pavlov's view of inhibition (adhered to by later Russian physiologists) reflects the present approach to the unity of subjectivity and physiology. Pavlov identified three basic forms of inhibition in accordance with their conditions of origin: external inhibition, internal inhibition, and sleep inhibition. External inhibition is produced by some external source, such as a drug; internal inhibitions are produced by the natural functions of neurons; and sleep inhibition is the brain's activity when asleep. The question can be asked: are all three forms of inhibition relevant to the production of nonsymmetry? Peter Anokhin has suggested that internal inhibition and external inhibition have the same physiological mechanisms but differ in their origins. "The data at our disposal indicate that the difference in the two . . . forms of inhibition lies mainly in their *conditions of origin* while their physiological mechanisms and physiochemical natures are identical."[30] If this is the case, then in terms of the production of nonsymmetry we can treat both internal and external inhibition as the same, for it is the physiological and chemical mechanisms that are the focus of our attention in relation to the production of the relations of meaning, not the question of origins.

In relation to the third form of inhibition, that of sleep, Anokhin draws a distinction between the active effects of internal inhibition and the passive synchronization of the cortical cellular masses during sleep; "during sleep a kind of third state of cellular elements exists, coinciding with neither external nor internal inhibition. Such a number of contradictions shows that it is absolutely essential to study further the relationship of sleep and internal inhibition."[31] For the present discussion, sleep is seen as a level of consciousness reflective of, but not separate from, wakefulness. This means that if subjectivity is structured by the three relations of meaning, so then will dreams and sleep without dreams. The condition of inhibition and the relationship of nonsymmetry are therefore too restrictive to equate with this complex set of conditions that clearly manifest all the relations of a wakeful subjectivity. For these reasons, I judge that sleep can be ignored in terms of the equivalence of inhibition and nonsymmetry.

Anokhin suggested that the physiological architecture of internal inhibition is the effect of competing excitations in the central nervous system: "the origin of inhibition in the central nervous system is the *encounter of two excita-*

tions."[32] Such a description tends to lead to the conclusion that inhibition could be seen as another kind of excitation. But this result is specifically opposed by Pavlov and, later, Anokhin: "Pavlov refuted the possibility of identifying the process of excitation with that of inhibition, although he did consider them to be identical in origin."[33] Anokhin's statement here is significant in relation to our proposed isomorphism of excitation and inhibition with symmetry and nonsymmetry. For while symmetry and nonsymmetry are different one from another, they do share the same basis or "origin": that of symmetry. In addition, the priority of symmetry is reflected in the same priority being given to excitation. Finally, the above description of inhibitions shows an integration of excitation and inhibition that is again isomorphic with the integration of symmetry and nonsymmetry. Inhibition and excitation were therefore perceived by Pavlov and Anokhin in terms almost identical to the description presented here of the semantic relations of symmetry and nonsymmetry.

It is unnecessary to develop a chemical analysis of the mechanisms of inhibition in order to propose the case for inhibition producing the semantic relations of nonsymmetry. Different chemicals have different inhibiting effects on neurons, but this does not affect the unity of meaning and neurology proposed here. This unity implies that despite the type of inhibition (whether presynaptic, direct, or inhibition from interneurons), a relationship will be produced by each one of these inhibiting mechanisms. If our initial proposition, that excitation produces the relationship of symmetry, is correct, the relationship produced by inhibition will be the relation of nonsymmetry. Why is this? When a relation of symmetry is impinged upon, modified, or negated to some degree (by the mechanism of inhibition), it is transformed into a more complex relationship. That more complex relationship will be one of nonsymmetry: a relation that contains both the initial relation of symmetry and the nonsymmetrical effects of some form of inhibition. In other words, if inhibition is the affect of the encounter of two excitations, then nonsymmetry is the affect of the encounter of two relations of symmetry. Such a description is entirely appropriate for an isomorphic view of nonsymmetry, for it fits the principle of emergence (referred to in chapter 3) in which the interaction of two symmetries will produce a nonsymmetry.

The Interaction of Symmetry and Nonsymmetry

From an isomorphic view, inhibition has a organizational integrity similar to the role that nonsymmetry plays in the construction of meaning and consciousness. As a physiologist, Anokhin viewed inhibition as a method the organism has of eliminating millions of possible movements in order to perform one single precise, coordinated movement: "inhibition must be regarded as a tool for the liberation of the organism from excess degrees of freedom that are potentially at its disposal at any given moment."[34] Anokhin's "freedom" can be seen as a reference to unlimited symmetry or excitation, and so the restricting role he gives to inhibition is entirely in agreement with the semantic role nonsymmetry plays in the various forms of consciousness. For instance, the selection and precision of any coordinated movement by the body can be seen in terms of a complex set of integrated processes of differentiation (inhibition). In other words, the motor neurons that produce movement can be seen as also producing a complex integrated structure of meaning at a prereflective level. Thus the integrating function of a movement is brought about by excitation (symmetry) in the form of reflex arcs and interneuronal loops, and the actions that bring precision to a movement are the nonsymmetrical relations (differentiation from inhibiting neurons).

These two semantic relations function reciprocally, operating by closed loops of different kinds (recursive closure) in order to perform a physical act with precision. For instance, on a voluntary level, random or uncoordinated movements (from too much generalized symmetry/excitation) are refined by repetition through a series of inhibiting processes that lead to finer and finer distinctions/differences of performance. This kind of equivalence is described in traditional terms by McLeish, who says, "differentiation of response, as a consequence of discriminating between stimuli, is based on the dynamic relationship between these polar opposites, excitation and inhibition."[35] This description automatically implies the unity of semantics and physiology explicitly proposed here. The only difference is that excitation and inhibition are not "polar opposites" but operate as a series of integrated and recursive processes that not only produce physical movement but, in a more complex manner, can produce the representational and reflective forms of consciousness.

On this general aspect of inhibition Anokhin draws our attention to its generalized applicability. For example, any coordinated movement always involves discrete and localized excitation, along with inhibition that extends down to individual synaptic formations and a general inhibition that integrates central nervous activity and behavior. This means that the overall operations of the nervous system are integrated and that inhibition can simultaneously be local and discrete and also operate at the level of the system in general of subjectivity. This is a complex and recursive concept of inhibition that makes it both discrete and local, yet a generalized function as well. The extent of the "general" but "local" operations of inhibition is isomorphic with the character of meaning and consciousness that I have been proposing.

Here again the language of physiology refers to the same functional processes as the semantic references used in this semantic theory. The difference is that traditionally in physiology the materialist emphasis has been on inhibition and its generality, while for the present perspective the focus is on the integration of differences within asymmetrical, asynordinate systems. From the present perspective it is the element of excitation that produces the integrating factor associated with discrimination, while it is inhibition that produces the effect of discrimination. Discrimination, however, cannot be separated from integration, and in the same manner inhibition cannot be separated from excitation or nonsymmetry from symmetry. These are interlocking, integrated processes, acknowledged by the traditions of physiology, but they are sometimes spoken of as binary opposites or at other times with an emphasis on the function of inhibition.

Anokhin, like Pavlov before him, was an exponent of psychophysical monism and so viewed general inhibition as one of the important components in that "still in many respects mysterious process, which has been called decision-making."[36] Indeed, from the present perspective the nonsymmetry of discrimination is a central aspect of all decision making, whether it involves the selection and combination of linguistic elements in the production of a text or the discriminations necessary for a coordinated movement.

It is also interesting to note the character of what Anokhin calls the "intimate physiological mechanisms." Anokhin is referring here to the effects of a range of different kinds of inhibitory influences. Inhibition can operate under varied conditions and influences, such as frequency and strength of excitation. But given these various influences on inhibition, there is no reason to suppose that these are inimical to the production of nonsymmetrical relations, quite the reverse. As I have said earlier, nonsymmetry should not be interpreted as the simple negation of symmetry. Such an interpretation would imply a complete blocking of all excitation and a complete nullifying of symmetry, which is not the case for either physiology or meaning. Rather, the term "inhibitory influences" implies a diversity of intensities of inhibition. Yet another description could be a "sliding scale" in complexity of inhibition. From the present integrating perspective such concepts imply the notion of degrees of discrimination that are associated with different forms of closure, such as units, unities, contexts, and constituents that we discussed in previous chapters.

It will be recalled that the concept of reflexive differences and recursive symmetries represents a circular system of meaning from which we can form the explicit constituents of conscious awareness. Under the present circumstances the concept of degrees of reflexive difference can represent the same physiological conditions as the notion of intensities of inhibition. From a physiological perspective the two sets of processes (excitation and inhibition) are always integrated processes. The same can be said of the semantic relations of symmetry and nonsymmetry.

By combining the language of semantics with physiology we can hypothesize that a range or scale of intensities of inhibition, due to the variation in the physiological and chemical conditions of inhibition, will display an equivalent range of discriminations that will accompany different forms of closure and the organization of subjectivity. From this perspective such a range or scale of inhibition intensities suggests the possibility of a range of organizations that will differ in terms of their intensity and location. In broad terms there could be at least three possible forms of organizations, depending on intensity of inhibition as well as on the location in the nervous system. For example, a series of relatively low inhibitory effects would tend to suggest relatively invariant prereflective structures within the nervous system at the involuntary level. Such invariant neuronal structures would seem to be entirely relevant to the multitude of involuntary actions that operate to regulate the body in a multitude of ways.

A second broad organization could be represented by the notion of a balanced or medium intensity of inhibition, the kind that would involve, in some complex capacity, the higher nervous centers as well as sensory and motor stimulation. These large circular structures would tend to indicate a relatively open recursive organization where cortical inhibition is moderate and where modification of behavior, learning, and the production of creative symmetry transformations could be generated. This kind of medium intensity of inhibition involving areas of the whole nervous system would represent the kind of isomorphic consciousness that is pertinent to the processes of experiential learning and that is also common to a holistic philosophy and spiritual development.

A third broad category of organization could be due to a high intensity of inhibition that would be manifest in the higher nervous centers of the brain. This kind of intense interneuron inhibition could produce local and specialized circular inhibition of inhibition. Semantically, this is reflexive differentiation and is a common feature of a rational orientation in which differences and discriminations are considered paramount. As I have suggested previously, this kind of intense cerebral inhibition, in extreme forms, would logically produce fragmentation and schizoid forms of representation. It would also produce rigid forms of representation and a tendency to ignore or erase subjectivity in mechanical interpretations. Where the first, relatively less inhibited circular structures are invariant and nonrepresentational and prereflective, the third would tend toward being rigidly representational.

It would be possible to speculate on a range of these semantic/neurological conditions that produce repression and different form of mental illness. This would be possible because of the proposed isomorphism of excitation with symmetry and inhibition with nonsymmetry. But I do not intend to pursue the mapping of mental illness here. Thus while the conditions that produce a variation in the intensity of inhibition are important physiologically, from a semantic point of view this variation has an important correspondence with various forms of perspective paradigms.

Views of Consciousness

How we view consciousness is directly relevant to the hypothesized unity of meaning and neurology presented here. For example, to return to the Morse code metaphor for a moment, a particular image of consciousness is implied when we use a mechanical language to describe the action potentials of neurons. If we say they operate to code and decode information, then inherent in this terminology is the consideration that neurons send messages and codes that are similar or the same as the codes of language that we use in all social interaction. But this assumes a fully reflective subjectivity that has the interpretative ability to code and decode a language at the level of the neuron.

But even if the physiologist or cognitive scientists reply that they do not intend to imply a fully reflective consciousness and that the codes are not language codes but chemical or electrical codes, then the difficulty is not less. Codes, whether chemical or linguistic, can be decoded and carry information. *Information* is an instrumental term, but it is not a useful term when it comes to referring to subjectivity or the operations of the nervous system. *Information* implies meaning and thus some kind of subjectivity. A subjectivity that is implied but never articulated is one that is traditional, and in the West this means a Cartesian cogito.

Terms like code, decode, cipher, decipher, information, calculation, and representation may be useful shorthand forms for the materialist to refer to brain activity, but these terms have semantic feet firmly planted in Cartesian soil. It is not possible to use exclusive terms for precise mechanical operations for anything. This maybe the materialist's desire, but it is nothing more than that, for there is no such thing as an exclusive and precise term. The problem with "hard-edged" terms like information, coding, and decoding is not that they imply more than they should. The problem is with the kind of Cartesian assumptions that are included innocently whenever they are used. Their use therefore establishes and reinforces the sight lines of a rational perception, one that is predisposed to always divide "mind" from body.

A rational, all-or-nothing, indivisible consciousness, the kind that Descartes is said to have proposed,[37] the self-aware kind that Eccles and Popper took for granted, is the consciousness of the rational "observer" that informs traditional academic study, as well as Derrida's deconstruction practices, and that is utterly given over to explicit meaning. This separate, rational, fully conscious awareness is rejected here in favor of an emergent form of subjectivity based upon the unconscious of symmetry. The concept of the unconscious of symmetry is the concept of an unconscious subjectivity. This is a dialectical image that links up with Merleau-Ponty's radical concept of the "body-subject."[38] Merleau-Ponty's phenomenology rejects the Cartesian view and argues that the "perceiver is not a pure thinker but a body-subject, and that any act of reflection is based on the pre-personal anonymous consciousness that is incarnate subjectivity."[39] From this point of view there can be no sphere of absolute self-sustaining thought implicit in the concept of the Cartesian cogito.

Merleau-Ponty has argued that the sensations of the flesh disclose a primary consciousness, a prereflective consciousness that is neither mechanical nor meaningless nor nonexistent. Nor is this prereflective, anonymous subjectivity separate from the reflective awareness but provides the conditions whereby a fully unified subjectivity is brought about.[40] This prereflective, anonymous consciousness of the flesh, this "incarnate subjectivity," represents for Merleau-Ponty the foundation of all empirical truth and certainty in a domain that is contextualized, human, and embodied. In Merleau-Ponty's view, experience is neither mechanistically determined nor a fortuitous construction, and our various explicit relationships with the world will always be subtended by a primordial background that itself cannot ever become entirely explicit. The implicit therefore always underpins the explicit in the sense that unconsciousness always subtends conscious awareness. Merleau-Ponty's phenomenological approach to consciousness, which links a prereflective consciousness to a fully reflective subjectivity, is therefore isomorphic with the present view of subjectivity as an asynordinate system.

Combining Merleau-Ponty's terminology with the present approach to meaning, it can be said that the system in general of subjectivity extends throughout the whole body, having levels of complexity that begin with a prereflective subjectivity that arises through the stimulation of the sensory and motor neurons of the nervous system. This consciousness of the flesh, which is constituted by the relations of symmetry and nonsymmetry and asymmetry, structurally underpins a fully conscious awareness that is established by the integration of the higher centers of the brain with the lower. It follows, then, that insofar as sensory and motor neurons are never isolated from the central nervous system or the brain, prereflective consciousness can never be separated from a fully unified reflective conscious awareness. Similarly, in the way that local and discrete excitation and inhibition can never be separated from the integrating influence of general inhibition, so incarnate subjectivity of the lived body can never be separated from self-reflective consciousness.

The Reflex

If we accept the existence of a prereflective consciousness, neural excitation and inhibition cannot be represented as the transmission of information. The so-called information of excitation and inhibition, for instance in reflex mechanisms, is the system of meanings generated within Merleau-Ponty's prereflective consciousness, and as such there is no passive transmission of already ordered code; rather at this local level there is the groundwork for the development of fully integrated subjectivity.

This kind of neurosemantic description and terminology can be extended to embrace even a physiological description of a withdrawal reflex: "A withdrawal reflex results in a rapid withdrawal (latency about 30ms) of a limb when a part of it is subjected to a noxious stimulus."[41] The pathway of the withdrawal reflex is comprised of sensory (afferent) neurons, a series of excitatory and inhibitory interneurons, motor (efferent) neurons, and skeletal muscle fibers.[42] In this oversimplified and abstract model there is already the production of prereflective meanings that is mostly unconscious but that has some elements of conscious awareness. For example, inhibitory interneurons provide the contact with the central nervous system and therefore the inhibition of the reflex. Karl Pribram has said that the reflex arc as a closed system is a fiction and that the mechanisms are directly controlled by the central nervous system.[43] We can thus say that the inhibitory interneurons together with the general inhibitions of the controlling central nervous system produce the nonsymmetries of inhibitions that are the meanings necessary for some awareness associated with the withdrawal reflex (pain, in this case).

It has been consistently argued throughout this book that the relation of nonsymmetry is *the* structural constituent of awareness. This means that the prereflective consciousness of the withdrawal reflex will always have some kind of component of awareness, because inhibition/nonsymmetry will always be involved. In other words, the withdrawal of the limb from the painful stimulus will be carried out largely unconsciously and involuntarily, but not entirely so, for the sensation and pain will involve an awareness of the pain by the organism.

A second point to be noted in respect to the withdrawal reflex is that awareness of pain will not be a fully reflective consciousness. If it were, then the pain in the foot could be distinguished from the registration of pain in the brain. In other words, there would be no symmetry of limb and brain as there is in all "normal" pain. This is not to say that we are unable to tell if we have stubbed our toe when we do so, only that the location of pain is not a matter of differentiation of space and time, but of the necessary symmetry of space and time, of brain and toe. The kind of subjectivity that is involved in pain from a stubbed toe will therefore structurally consist of a symmetry (unconsciousness) of space and time. In other words, there will be no differentiation of space and time between the initial stimulus that initiated the withdrawal reflex and the registration of the ensuing pain, even though these points may be more than a meter apart. While this prereflective consciousness does not differentiate between the objective factors of an "observed" space and time, the prereflective structure of this subjectivity nevertheless operates on a background to figure-structure. We could therefore say that awareness, produced by inhibition (nonsymmetry), will stand out as pain against the background of unconsciousness (excitation/symmetry).

Implications for Conditioning

By the simple expedient of transposing the terms symmetry and nonsymmetry for excitation and inhibition we are able to give a semantic and subjective reading to any neurological activity. If this is so for the withdrawal reflex, it is also possible with the conditioned reflex in dogs. Pavlov found that in dogs a conditioned salivation was established by repeatedly pairing some arbitrary stimulus with food. In the process of establishing a conditional response, there is characteristically a generalization of response from the particular conditional stimulus to others that resemble it. For instance, if a tone of 1,000 cycles per second is established as a conditional stimulus, other tones of close frequency spontaneously acquire similar properties. In addition, the ability of other frequencies to evoke the conditioned reflex diminishes in proportion to the interval between these frequencies and 1,000 cycles per second. Similarly, if touching a definite location of skin is made into a conditional stimulus, touching other areas of skin will evoke the conditional reaction, but the effect diminishes with increasing distance from the original location. This effect of similar stimuli evoking the same conditional response is called "stimulus generalisation."[44] It works on the basis that the greater the similarity of stimulus, the greater the stimulus generalization.

It would be impossible for conditioning to work if it operated solely on the basis of the traditional and explicit behavioral terms used here to describe it. These instrumental terms as traditionally used are far too mechanical to suggest reasons why an arbitrary stimulus such as a tone of 1,000 cycles per second would evoke salivation. Nor can the terminology of "stimulus response mechanism" suggest reasons for stimulus generalization. There is nothing in the use of this terminology that explicitly addresses what is undoubtedly implied, that is, the symmetry of the two elements essential for conditioning. For example, what is there in this traditional language that suggests an answer to why a tone of 1,000 cycles per second should be symmetrical with food? The question of meaning and semantic structure was never addressed by Pavlov or by behavioral psychologists. Nor is there anything in this traditional terminology that suggests why stimulus generalization should work. Why should two similar tones evoke the same response when the tones are different? Clearly there is another factor at work. And this other factor is the meaning of symmetry produced by excitation.

For instance, in the case of the repetition of arbitrary stimulus and food there is a doubling of symmetry that, when repeated often enough, begins to operate as a closed recursive system. The closure of this prereflective transient system is brought about by the symmetry (excitation) of interconnecting nerves. Closures, whether transient or invariant, are always the function of symmetry circularity. If any of the sensory neurons involved in this system are stimulated, their excitation will automatically tend to involve the other neurons in this system. But the generalized excitation of this system will tend to be restricted if the arbitrary stimulus is repeated without the presence of food. This will be so because inhibition (nonsymmetry) is always part of excitation, and thus in any system of excitation there will also be the inhibiting influences of nonsymmetry at work. The effect of inhibition (nonsymmetry) on excitation (symmetry) appears delayed by some factor of time that will depend on the relative strength of symmetries (for instance, the similarity of stimulus). The term "similarity of stimulus" refers of course to the notion of symmetry inherent in the stimulation of neurons. This is the case with stimulus generalization, in which stimuli similarity is high. Given sufficient repetition, however, stimulus differentiation (nonsymmetry) will be brought about because the effect of inhibiting neurons will, over time, reduce the number of pathways to a minimum.

I suggest that without some implicit or explicit notion of symmetry and nonsymmetry it is impossible to make any sense of conditioning, or any other neurological activity for that matter. This proposal implies that such classical terms as "stimulus response mechanisms" should perhaps be abandoned for more subtle and complex descriptions that come from the integration of symmetry, nonsymmetry, and asymmetry.

Pavlov's experiments in conditioning were carried out on dogs. The level of subjectivity with which he was dealing was, then, very much below the level of reflective conscious awareness. It is no wonder, therefore, that behavioral psychologists, in transferring this model of learning to humans, found so much criticism of their methods. The level of semantic complexity of a reflective awareness that deals in representations is so much greater than a subjectivity that does not operate with representation. Because of this greater complexity there are dangers in transferring the system of conditioning of dogs to humans. When, for example, a so-called spanking stimulus is administered to a child, it will affect many more levels of excitation (symmetries) than those simply and directly connected to some specific, unwanted action. Even in dogs, relative directness of connection and simplicity between a series of stimuli and a response is difficult to control. Even more so in humans, in whom the recursive complexity of a reflective consciousness has systems recursively coupled to systems, coupled to systems, et cetera. In this case the effects from a set of stimuli, for instance "spanking," can reverberate throughout the whole psychic formation. The symmetries associated with self-images will automatically be involved because of the general integration of the whole nervous system. Negative conditional responses will therefore entail not only the reduction of an unwanted behavioral pattern but, in addition, negative self-images. Conditioning as a learning instrument may be applicable to the nonrepresentational forms of subjectivity in dogs, but for humans it is totally unreliable and inappropriate, except for the most negative of purposes.

The "Body Image"

I turn now specifically to the notion of the prereflective consciousness that produces what Merleau-Ponty calls our "body image." The view that subjectivity has several levels of operation is not foreign to contemporary physiology, which uses terms like "coma," "semicoma," "stupor" and "confusion" to refer generally to various levels of consciousness. In addition, in the literature on neuroanatomy we can find such scales of consciousness as the Glasgow Coma Scale.[45] This scale is a method of assessing changes in the levels of consciousness based upon eye opening and other motor responses, as well as verbal responses. This physiological view of the different levels and intensities of consciousness reflects to some degree Merleau-Ponty's emergent view of a prereflective consciousness that subtends reflective consciousness.

If the physical body has a consciousness—a prereflective consciousness—then it should have some kind of body awareness in the form of a body image. In this respect Merleau-Ponty uses the concept of the "lived body" that has a body image. A lived body is, however, a different type of object from other objects. Merleau-Ponty makes the important distinction between objects of the world, which are seen to have the coordinates of time and space, and our bodies as objects, which do not have these "objective" factors. From introspection we know that the parts of our body are not related to one another as objects occupying a place in objective space. As Monika Langer says, "The body is immediately present to us because we *are* our body."[46] Thus we know where our limbs are without having to look at them because we possess a body image. According to Langer, our body image provides us with a prereflective knowledge of the location of limbs, but this location is not a position in objective space: "Rather, it is a location with reference to the way in that my limbs enter into my projects; thus it is not a 'spatiality of position but a spatiality of situation.'"[47]

This spatiality of situation implies a body as experienced rather than the body as differential object, and this distinction shows up dramatically in body images that include phantom limbs. Psychological investigations of phantom limbs show that a body image can include a limb that has become actually nonexistent. How is this possible? Merleau-Ponty suggests the concept of "incarnate intentionality." Incarnate intentionality refers to some kind of awareness of the body by the body; it is a knowledge produced at a prereflective level of consciousness. As Merleau-Ponty fails to give an account of meaning, he neglects to describe the subjective structure of incarnate intentionality. From the present perspective the meanings inherent in incarnate intentionality are structurally the same as in the withdrawal reflex. These are the relations of symmetry and nonsymmetry and asymmetry produced by the excitation and inhibition of the system of neurons involved.

In terms of the meaning of a phantom limb, we can give a similar reading. If we already assume that the large sensory axons of the limbs produce an overall symmetry of relationships between receptors and the central nervous system because such excitation is largely uninhibited, then it is a small step to realize that this inherent nondifferentiation of meaning could figure in the body image that produces the phantom limb. If a limb no longer exists, but some part of the sensory axons of the limb is still operative, then excitation of these severed nerves should reproduce the images of a fully functional limb. This is the case because the meaning of the prereflective consciousness that produces the body-image is one based on the symmetry of time, space, and location. In other words, if the nerves in the limb stump are stimulated in any way, the meanings so produced will be those that reconstruct the missing limb. This will be the case because a prereflective consciousness will not take into account the differential conditions of location or time and space, factors related to the severing of the limb. Nor will a prereflective consciousness take into account the differential reduction of the length of limb. Given the absence of the differential conditions of space and time and length, the now missing limb will be erased because at this level of subjectivity there is no ability to make these kinds of distinctions. If these objective factors are missing along with the limb, then through the symmetry of habit and the symmetry of excitation the image of a fully functional limb should be reproduced.

Only a fully reflective consciousness can make distinctions between space and time, but this complex level of differentiation can never be achieved with one's own body because, as Merleau-Ponty points out, our bodies are different from other objects. Our bodies have a spatiality of situation, not a spatiality of position (as external objects appear to have). Therefore the prereflective consciousness that always subtends our fully reflective consciousness will affect our comprehension of our body image and body sensations by omitting such differential factors as space and time. For these reasons space and time are factors generated by a fully reflective conscious awareness that can only situate objects other than our bodies.

When stimulated, the stump of a severed limb, by itself, can produce no reflective consciousness, which means that the speed of the impulse is ignored. Speed and time are therefore simply hidden factors that will of necessity be masked in the very construction of this symmetrical relationship. In terms of survival, these "immediate" prereflecting conditions of symmetry, which operate the body unconsciously, are normally necessary for the smooth coordination of involuntary reactions in order for the body to respond quickly to a changing environment. It is only under abnormal conditions such as the loss of a limb that such symmetries appear incongruous.

What should our view be of these relations of symmetry that hide the objective and rational factors of space, time, speed, and frequency of impulses? Is the effect of a phantom limb simply a case of rational misrecognition? The symmetry of time and space produced by neuron excitation could be considered as misrecognition only if the structure of meaning is ignored and the legitimate viewpoint shifted onto the detached "observer," that is, another subjectivity. But to call another's pain or sensation a "misrecognition" is too prescriptive to contemplate. The point being made here is that the phantom limb is a "natural" effect of an abnormal situation. It is not an abnormal effect of a rational situation.

From the present isomorphic perspective, when a toe is painfully stubbed, even though the two points of contact and painful effect are far apart in space and some milliseconds in time, a healthy, "normal" nervous system operates on the basis of the symmetry of cause and effect. This unity of pain and site of stimulation is the same unity that is established between the end of a limb and the connection it has with the central nervous system. Because of these symmetries, the prereflective consciousness of sensory stimulation does not make objective space and time part of the body image when a limb is severed. For these reasons the symmetrical structure of a body image can include the semantic structure of a phantom limb.

The Split Brain

The implicate order by which the body communicates with itself is quite evident in what we know of split-brain experiments. When the two hemispheres of the brain are cut by severing the corpus callosum, patients experience two "consciousnesses," one located in the right side of the brain and one in the left. While these "consciousnesses" are different one from the other, it is still possible to communicate separately with each hemisphere, as the well-known studies of Roger Sperry have illustrated.[48]

What can these split-brain experiments tell us about meaning? The results of split-brain operations have some semantic characteristics similar to the experience of phantom limbs. Like the phantom limb, split brains produce severed neurons and therefore severed symmetrical relations, with similar resulting incongruities. Whereas the effects of a severed limb draws our attention to the symmetrical integration of space and time, a severed corpus callosum points up the normal integration that this fiber bundle brings to the two hemispheres. With the limb the integration is simpler, and so the cutting off of a limb produces simpler incongruities. With an uncut corpus callosum the two hemispheres of a brain when stimulated produce an integrated, reflective conscious awareness. When these two hemispheres are severed, this physical division of the brain is identical to the division that has taken place in consciousness. The division of conscious awareness therefore corresponds to a split brain.

Now this physical division is also a division of meaning; it is a division of the meaning of symmetry and an integrating order. When a symmetrical relationship is divided "unnaturally" by the cutting of neurons, such severing does not produce an inhibition or a relation of nonsymmetry or a sense of difference or an awareness of separateness. Each of these differential factors is only produced by intact inhibiting neurons working in harmony at a high level of recursive complexity. A split brain does not produce increased inhibiting neurons, only cut ones. When the two hemispheres are divided, the normal integration of conscious awareness cannot, therefore, take place, but because such cutting does not develop any inhibiting mechanism in relation to the division, there is no awareness by the patient of their two separate consciousnesses. The symmetries normally produced by neural excitation are no longer, and the usual integration of conscious awareness is therefore fractured.

Thus the effect of the severing of the brain is similar to that of a lost limb. The meaning of normal symmetry is interrupted, resulting in either a phantom limb or two sites of conscious awareness. In both cases, however, the simple and the complex, the incongruities produced by severing are the incongruities of the normal semantic effects of symmetry in the operations of subjectivity, operations that are structured by the meaning of symmetry.

Conscious Awareness

What do these interpretations tell us about the nervous system? Surely that the normal nervous system produces a subjectivity structured by the relations of symmetry, nonsymmetry, and asymmetry: integrated relations that function in varying degrees of recursive complexity, forming prereflective and reflective fractals of consciousness. Certainly this approach denies the traditional and rational view that consciousness resides in a single location (the so-called dominant, or left, hemisphere).[49]

We come then to the physiological nature of conscious awareness, to the traditional question asked about the unity of mind and body. To a large extent the isomorphic approach taken here has already answered this question. But this answer has not been the traditional kind. We have not, for example, considered the mind as a closed, featureless awareness of intentionality, in the Popper, Eccles, and Searle mould, and therefore I am not suggesting the liaison areas of the left hemisphere as the probable site of a rational consciousness.[50] As I have already indicated, this perspective of reason in relation to subjectivity is a waste of time and will not provide a satisfactory answer to the unity of subjectivity and body.

From an isomorphic perspective it has already been indicated that if meaning is generated by neural excitation and inhibition, the whole body will participate in subjectivity. The unity of physiology and subjectivity will therefore be one that involves the entire body, which generates varying degrees of recursive complexity, from prereflective consciousness to self-reflective conscious awareness. It is therefore not a question of a simple one-to-one relationship, involving a specific brain area with a separate, featureless cogito. Rather, the question concerns an equivalence of physiology and semantics in terms of levels of complexity and organization.

As a general comment, if neurons produce semantic relationships as well as physiochemical reactions, then conscious awareness must represent a high level of recursive complexity, that is, a high level of neural excitability. This must be the case because conscious awareness represents a high level of recursive asymmetrical complexity, such as is evident in the meaning of a four-dimensional manifold of space/time or within the subtext implications of a particular discourse. From this standpoint conscious awareness is the result of a high level of recursive neural excitation and inhibition, which probably involves large and diverse areas of the human nervous system.

In line with this view that conscious awareness is generated by a complex, general, and integrating set of neural pathways is the knowledge, collected by brain surgeons and neurologists, that the higher brain stem (the reticular activating system, RAS) is the indispensable organizer of consciousness. The Canadian neurosurgeon Wilder Penfield has emphasized that the RAS is not a center of consciousness but rather acts as a coordinator and unifier of the activities of the higher and lower center that are involved in the generation of consciousness.[51] Derek Denton tells us that the neurons of the RAS are structurally unusual in that one branch of cells "descends down the brain stem while the other goes upwards to the cortex."[52] Effects of this process can fan out over the entire cortex, thereby integrating into a complex system the lower and higher centers of the nervous system.

This is exactly the kind of neural network one would look for in order to find the necessary integrated recursive structure suitable for reflective conscious awareness. The coordinating role of the RAS is exactly the integrating role needed to account for a complex system involving both prereflective and reflective systems. Hemisphere location is therefore not as important for conscious awareness as those neurological structures that provide the integration, unification, and coordination of recursively complex systems.

It is interesting to compare how Pavlov saw consciousness in the 1920s with the kind of integrating descriptions Penfield makes about the RAS. While

Pavlov is ambiguous about the difference between a human and a dog's brain in relation to subjectivity, he nevertheless has this to say when he relates consciousness to brain activity:

> consciousness appears as a nervous activity of a certain part of the cerebral hemisphere, possessing at the given moment under the present conditions a certain optimal (probably moderate) excitability. At the same time all the remaining parts of the hemispheres are in a state of more or less diminished excitability. In the region of the brain where there is optimal excitability, new conditioned reflexes are easily formed, and differentiation is successfully developed.[53]

Pavlov's view of subjectivity is interesting, for he employs an image of an ever-changing center of excitability surrounded by outlying areas of lesser irritability. Such an image reflects to some degree the kind of complex, encompassing structure proposed in this isomorphic theory. Pavlov believed that the area of relatively high excitability was the most aware and creative part of the brain. By this he meant that differentiation (inhibition) was most highly developed, where new meanings are formed and older ones transformed. Pavlov's description also suggests that conscious awareness in humans comes from a certain optimal excitability that migrates over the whole extent of the hemispheres and involves lower as well as higher centers. The interest of this description is that he could not possibly be referring to the model of a single, featureless Cartesian cogito. Rather, Pavlov's description foreshadows the present emergent view of consciousness, one in which the whole of the nervous system is involved in the structure of subjectivity.

Conclusion

The present thesis for the unity of material body and subjectivity is clearly not a dualist proposition. The body is here not considered to be a machine and the mind is not viewed as a parallel or separate site of pure intentionality. Nor is the present approach a physicalist, reductionist thesis, one that proposes that subjectivity is simply what the brain does. The present isomorphic approach proposes, first, that subjectivity is a complexity of relationships that emerge from the semantic relations of symmetry, nonsymmetry, and asymmetry; second, that these three basic relations are recreated within subjectivity by neurons when in the states of excitation and inhibition; third, that the complexity of reflective consciousness most probably arises through the organizing role of the reticular activating system.

A subjectivity structured by semantic relations implies an incarnate consciousness of the body, a prereflective consciousness that subtends conscious awareness. This is a structure in which the explicit emerges from the foundation of the implicit. This kind of architecture fits the architecture of the nervous system through the functioning of the mechanisms of inhibition and excitation that from a semantic point of view are the relations of symmetry and nonsymmetry. Such is the basis for the unity of subjectivity and material body.

Part 3

The Context of Discourse

Chapter 10

The Isomorphic Sign

The discursive context of subjectivity represents the third "layer" of consciousness that emerges out of the context of the system in general of subjectivity, which in turn has emerged out of the context of the symmetries of cosmic consciousness. This chapter begins to develop an isomorphic reading of discourse and compares this interpretation with the traditional semiotics of reason.

The discursive dimension of subjectivity is significant because it is through this "second signal system" that we can reflect on and inscribe the world. Discourse represents the means whereby subjectivity is made manifest, social, and self-reflective. Discourse provides the communicative paths by which we traverse and explore the terrain of our own subjectivity. These paths are not innocent, opaque, or free-flowing but are self-reinforcing sight lines that often restrict and constrain as much as they liberate. The closures of our discourse tend to be the closures of our reflective processes, and discourse is a system full of closures.

Appropriateness

Often in this book I have stressed that this or that form of terminology is inappropriate to represent the complexity or the inherent integration of a problem at hand. Outside the constraints of conventional etiquette, the idea that some kinds of phrases are inappropriate is novel. It is novel to the view that consciousness is a featureless unified state separate from the world and that the world is a state to be represented using the formal tools of language. The link between "mind" and the world is seen to be established through the use of the social tools of language. Language tools are like other tools in that they need some practice and skill to use appropriately, but beyond the etiquette of literacy, they are not differentiated. Thus if language tools are social and beyond individual control, appropriateness will be related to social rules of spelling, grammar, or genre and not to the perspective frameworks of individuals.

This is an explicit example of the implicit reasoning of many rational linguists. This view is challenged here with an isomorphic view of discourse. Such a view begins by saying that while the morphic field of meaning is universal and that nothing exists outside it, nevertheless within subjectivity there are three paradigms of consciousness. As discourse is always constructed within the limits of subjectivity, it will not only be entirely subjective but also will reproduce a perspective framework. The sentences we speak or write are therefore not formal, innocent tools but vibrant constructions that reflect and reinforce the values of the underlying perspective.

Appropriateness is therefore a question that goes beyond etiquette to perspective paradigms. Certain phrases and the use of some terms can be inappropriate because that which they seek to represent has characteristics beyond the limits of the phraseology's paradigm. This is often the case with the paradigm of reason and its linguistic conventions. Paradigms of consciousness therefore do not exist only in nonverbal psychological states but are supported, produced, and reproduced through the resonance of their own discourses. Thus rational language will reproduce rational thought, and isomorphic discourse will reproduce an isomorphic perspective. In rational terms it is impossible to separate language from thought. In isomorphic terms it is impossible to separate discourse from perspective.

Linguistic Conventions

What are some of the linguistic conventions of reason? This is a significant question for any research that seeks to discover the links and relationships of any particular system, whether quantum mechanics or social formations. It is significant because if the linguistic conventions of reason are inadvertently used, it will be impossible to adequately represent the holistic complexity of the system's relationships and syntheses. It will be impossible because reason's terminology will keep recreating a perspective of reason, one where differences are exclusive and integrations are either unmade or bracketed.

As previously stated, the linguistic conventions of reason deal with the principle of exclusive difference. In practice, how is this principle inculcated in discourse? This question involves how the generic of perspective is translated into specific linguistic use. As each specific case will be different in each discipline and in every area of each discipline, specific cases pose difficulties in explanation. Nonetheless, I intend to investigate in the next two chapters the specifics discourse itself. I have picked this area because it represents a third "layer" of consciousness (after the system in general of subjectivity and cosmic consciousness), but also because of the generic role that discourse plays in all social interactions.

Before we begin to explore an isomorphic reading of discourse, some cursory comments should be made about how the principle of exclusive difference is inculcated in the discourses we use. This question points to rhetorical strategies as an area of study. "Rhetorical strategies" refers to those linguistic strategies that seek to hide yet reinforce the underlying perspective and its accompanying set of values. Rhetorical strategies deal with the desires of the author and reader: their agenda and how they intend to achieve it, as well as the conventions and agendas of the social institutions with which authors and readers interact. All these private and social values are integrated and framed within a particular perspective framework within any piece of discourse. To study rhetorical strategies is therefore to uncover these several levels of hidden meanings.

Because our focus here is reason, we will not be dealing directly with specific authors, institutions, or conventions, except in passing. Reason has a series of conventions to ensure its perpetuation through linguistic reinforcement. A common one, and close to the present discussion, is the denial that problem or puzzle solving is affected by the cultural conventions of the discourses used. The scientific community's reaction against Thomas Kuhn's *The Structure of Scientific Revolutions* provides ample evidence of the need for this denial. A materialist and empiricist's view of things usually entails the bracketing out of discourse, that nonmaterialist area of study, from its calculations, even though calculations are always discursive. Kuhn's mild suggestion that the cultural conventions of a scientific community were of importance for scientific advancement was met, in some quarters, with total denial.

Why is it that discourse and rhetorical strategies seem so inimical to mechanical science? The answer is not in the emotional lives of individual scientists but in the perspective of reason they continually reconstruct. Discourse not only humanizes, relativizes, and makes puzzle solving more complex, it also deals directly with relations, and relations are always relations of meaning and therefore nonmaterial. For materialists to encompass the proposal that the nonmaterial is important means they will have to change their perspective paradigms. But entrenched identification patterns, reinforced by perspective sight lines that have been built up over years of training, are not easily changed. Denial is thus a common response to any serious challenge to reason.

Another of reason's linguistic conventions is the tendency to use words and phrases instrumentally, in the sense of "using an instrument." Instrumental linguistic practices tend to rely upon conventions of use without further explanation. This is the practice of stereotyping important structural elements within a conceptual system. Cognitive science is a field where instrumental language hangs like lights on a Christmas tree. Cognitive science is full of instrumental words and phrases that unfortunately are the bedrock of the discipline, phrases such as "encoded representation," "encoded information," "sensory coding," "decode the encoded forms," "discrete signals," "artificial intelligence," "physical symbols"—the list could go on and on.

Instrumental linguistic practices help to establish mechanical processes by erasing the possibilities of interactive transformations. These practices give an appearance of selectivity and precision through the denial of the implicit, and all that means in terms of knowledge, consciousness, and the world. Instrumental linguistic practices play to appearances, for surface readings are usually the only reading. The possibility of transformation from surface constituents to hidden context is therefore denied by these kinds of practices, with decontextualization the mark of an instrumental practice.

Disciplines that seem prone to the use of instrumental discourse are those where disciplinary identity is often in question and there is a consequent emphasis on the trappings and conventions of reason (psychology, cognitive science). These are the trappings of rigor, analytical reasoning, "objective" research, the extensive use of "facts," dualistic argument, the use of statistical analysis, and pedantry. In addition, often these disciplines are faced with problem solving in the very areas most unsuited to a mechanical and reductive approach. Such disciplines as education and linguistics spring to mind here, although the practice of instrumental use is pervasive throughout the academy and not least in the science areas. More of the rhetorical strategies of reason will be referred to in chapter 12.

The Isomorphic Sign

I would like now to return to the question of an isomorphic reading of discourse. This is a question that rereads established linguistic theories and adds to them the features of subjectivity and implicit meaning. As already noted, reason tends to erase subjectivity and the implicit from all interpretations. In traditional linguistics subjectivity tends to be relegated to the personal and idiosyncratic domain, which is seen as different and separate from the social, which is what language is supposed to be all about. Implicit meaning also tends to be erased from traditional linguistic theory by being regarded as explicit meaning that is not in use or that is not being attended to at any given moment.

In this manner *language* is a term used to refer to that sometimes material, sometimes formal and transcendental system, of linguistic forms. One of the most influential and fertile approaches to language this century has been Ferdinand de Saussure's structuralist theories. Saussure developed his theory of language based upon the structure of his formalist sign (signified over signifier):

Signified
Signifiers

Saussure's "sign" is classic reason. It lacks explicit reference to the material world, subjectivity, or the body. The existence of a body is, however, implied in Saussure's explanations of the speaking circuit.[1] Saussure's description of the speaking circuit refers to speech—*parole*—which is language *in use*. Speech or language in use is different from a formal language *available for use,* or, as some psycholinguists call it, "language knowledge."[2] As linguists know, Saussure produced no theory of *parole,* no theory of language *in use*. He was preoccupied with his sign structure and with "language" not in use, as an ideal Platonic formal entity, divorced from the generic context of subjective or specific contexts of the environment.

In contrast to Saussure's sign structure of language not in use, an isomorphic approach would redesign this structure to give it both generic qualities and the implication of specific contexts of application. An isomorphic sign structure for language *in use* will be for *discourse* and would need the additional structural element of the meanings of the prereflective, presymbolic body to subtend the other two levels of signifier and signified. Such an isomorphic model situates the prereflective meanings of incarnate subjectivity at the first level of discourse, prior to the levels of representation involving signifiers and signifieds. Without the prereflected meanings of subjectivity there can be no so-called language, for there would be no production of symbolic meaning. The category of the prereflective subjectivity therefore represents the primary condition of the isomorphic sign. The isomorphic sign, as will be recalled, has the following hierarchical structure:

3 Signifieds: conceptual structures
2 Signifiers: perceptual structures
1 Prereflective bodily meanings

The isomorphic sign explicitly refers to the three levels of subjectivity that are engaged when that subjectivity is involved in symbolic activity. This was also the implication within Saussure's sign, but it was never explicitly stated, nor could it have been for paradigmatic reasons. Because the isomorphic sign refers to levels of subjectivity, it is not a formalist or idealized Platonic structure but rather a psychosemantic-cultural model.

It should be remembered, however, that this sign structure does not provide a completely holistic image of meaning. It refers only to the levels of subjectivity that are involved in discourse and neglects the wider context of the relations of meaning in the domain of cosmic consciousness. To schematize this final context we would need at least two more levels, one prior to the first level and one after the third level and both referring to the symmetries of cosmic consciousness. Such a five-part configuration may provide an image of continuous and holistic transformations between all the levels, but the model would seem to be too unwieldy for a focus on discourse.

Given these qualifications, the question arises as to whether it is worthwhile having any isomorphic model of discourse. It might have been better to draw attention to the mechanistic implications in all models and to their limitations in representing a recursive and holistic system. Such a response appears overly negative, as it seems necessary to make a positive contribution with a set of proposals that go beyond the usual, traditional academic negation. For these reasons I consider the isomorphic-sign model worth proceeding with, even though it omits the final context of cosmic consciousness.

Given these qualifications, we can say that the isomorphic sign is a generic structure that refers to the conditions of subjectivity when it is manifest in its many cultural forms. From the present perspective a discourse is a recursive and integrated set of representational, rhetorical, and behavioral processes involving the three levels of the isomorphic sign.

It should be stressed that the spaces between the three levels should not be taken as absolute divisions or separations, for these differences represent a series of interrelated functions in a normally integrated human nervous system. For example, in speech the eardrums are initially involved in receiving vibrations, the meanings of which are then passed on along auditory pathways, by excitation and inhibition, to the Wernicke area of the brain, by which time signifiers (sound images) are constructed.[3] From there it appears that many other higher centers of the brain are involved in the emergence of conceptualization: the production of signifieds as concepts and images.[4]

The isomorphic sign assumes that the anonymous prereflective body is the primary and dominant feature in the construction of symbolic meaning, discourse, and culture. It is an individual author and reader who contextualizes, situates, motivates, and produces the discourses of culture. It is an individual subjectivity that provides viewpoint and coherence and constructs the rhetorical strategies and an aesthetic and political perspective in every "text." Specifically it is the immanence of prereflective meaning situated in the activities of eardrums, eyes, tongue, vocal cords, and fingers that provides the primary conditions for the production of all "discourses."

The process of perceiving a signifier (the consonant "t" for instance) happens "instantaneously" in that the hearer is unconscious of the different levels of the sign and is only conscious of the identity of the final image: the consonant "t." In other words, we will be aware of the percept "t" but not the several perceptual processes by which the "t" becomes a percept. The relationships between these first two levels of the sign are therefore integrated, transformational, and unconscious. The third level of signifieds is equally integrated into the sign structure, and the complexities of this relationship will be discussed in the next chapter. The distinctions within the sign structure should therefore not be taken to indicate separate and discrete levels of abstraction.

Recursive Integration

A further point should be noted in respect of the three levels of the isomorphic sign. That is, the hierarchy of these levels should not be taken as indicating a simple linear progression of level 1 through to level 3. Rather, each level is a transformation of the previous level, with the first level being a transformation from the symmetries of cosmic consciousness. This hierarchy is also an integrated and recursive system that functions as a "vertical" series of feedback circuits within feedback circuits. Thus level 1, composed of a complex series of recursive feedbacks, is linked to levels 2 and 3, also composed of complex series of recursive feedbacks through an additional series of recursive back circuits. And level 3 is linked to level 1 in the same manner. The term *feedback* is used here as equivalent to the concept of recursive symmetry. Thus the symmetry is recursively transformed at each of the three levels.

Recursive symmetry is characteristic of the isomorphic sign, as it is of all holistic systems. The physiology of the nervous system is hierarchically organized in that successive levels of interneuron produce successively more complex behavior and meaning.[5] And the isomorphic sign is hierarchically structured in that signifiers and signified emerge out of the prereflective meanings of the body. The hierarchy of neurological complexity is thus reflected in the hierarchical structure of the isomorphic sign. The meanings at each level of the sign are successively more complex than the preceding levels of meaning that they incorporate. The isomorphic sign therefore represents a complex and hierarchical system of meanings, produced by a hierarchically organized human nervous system, seemingly appropriate for the complex adaptive behavior needed for the survival of the species.

A further point of distinction between Saussure's sign and the isomorphic sign is that in the isomorphic sign the integrated coupling of the three levels is not *arbitrary* but *necessary*. The necessary connection of the three levels of the sign does not mean that whenever some aspect of prereflective consciousness is activated the other two levels are automatically involved. Most of the body's automatic and specialized systems are not automatically engaged in the production of discourse and symbolic activity. In terms of the sign structure, however, it is no use distinguishing between those bodily processes that are directly involved in communication and those that are not. Since the body functions on the basis of an integrated system of symmetrical relations, such a distinction, for the present purposes of discussing discourse, is an unnecessary diversion.

The connection of prereflective meanings to signifiers and then to signifieds is structurally produced only when the second level of the sign is constructed, that is, when some form of symbol is involved. When this happens, a signifier is automatically formed through a complex series of unconscious meanings produced by excitation and inhibition in an integrated series of areas of the nervous system. Once formed, a signifier will also always produce a signified (this controversial point is argued fully in chapter 12). All discourse will thereby be produced through this three-tiered structure. But the sign structure itself, while providing the necessary conditions for the production of discourse, only comes into operation in relation to the formation at the second level of some kind of signifier.

It should also be recognized that the three paradigms of consciousness, the symmetricized, the rational, and the isomorphic, do not equate with the three levels of the isomorphic sign. The three levels of the isomorphic sign represent an invariant structure through which all discourse and manifest consciousness is constructed, while the three paradigms of consciousness function as the three possible perspective frameworks that can relate to any system at all. Therefore a paradigm of consciousness can function with or against the "natural" bottom-up order of the isomorphic sign.

The concept of a reversal of the "natural" order of the isomorphic sign is important here. Anything can be viewed from each of the three orientations of consciousness. For example, a symmetricized perception of one's own body can be one in which repressed childhood desires, fears, or pains are displaced onto the adult body in a series of overdetermined and perhaps either negative or positive narcissistic and/or painful activities. Overly positive and negative examples of the unconsciousness of symmetricization toward one's own body (e.g., grandiosity and self-loathing) abound in psychoanalytical literature, and we need not go into them here. From a position of meaning, however, a negative symmetricized perception of the body (self-loathing) reverses the priority of the isomorphic sign by both negating the body and at the same time reifying it into some form of abstract negative representation. A positive symmetricized body (grandiosity) tends to be the flip side of the negative representation. Both however, are reified abstractions.

Next, a rational view of the prereflective body is to ignore it (as in most science, mathematics, linguistics, and philosophy) or represent it as a site of mistrusted desire (some forms of Christian and Cartesian theology) or view it as a mechanical machine, as in Western medicine. In each of these perceptions there is a reversal of the "natural" order of the isomorphic sign. There is therefore often a dissonance (not a resonance) associated with reason in respect to the body. In these rational discourses the body comes second in importance to the presumed primary symbols of a rational set of mathematical, logical, formal, or philosophic representations.[6] This reversal means that the second and third level of the sign are given priority over the prereflective and ineffable systems of meaning from which the second and third levels have emerged. This is exactly the position of the Saussurian sign.

Finally, the prereflective body can be perceived from an isomorphic perspective, a viewpoint that has been represented by varying degree by writers such as Martin Heidegger and Merleau-Ponty and in some of Alfred Korzybsky's writing. From this isomorphic perception the prereflective body has the same priority as it has in the isomorphic sign. It is placed in neither a negated nor a revered position; nor is it the condensed site of many discourses. Rather, it is the context for the production of meaning and the context in which discourse, communication, and culture take place. The body produces a field of meaning that reflects the implicate/explicate structure of the cosmos. It is this same asynordinate structure that we find in discourse.

Thus from an isomorphic perspective, the hierarchy of the isomorphic sign represents a "natural" order and the necessary condition for the construction of discourse and the cultural manifestation of subjectivity. This natural order is an cyclical order of context to constituents, with the prereflective meanings of the body as the context and the signifiers and signifieds as recursive constituents.

The Prereflective Body

We return to the first level of the isomorphic sign to explore some of the prereflected meanings that relate to discourse. This is the level at which the body involuntarily and automatically interacts with itself and the environment.

It will be recalled (chap. 4) that the environment is viewed as an energetic, asynordinate system having the same semantic structure as subjectivity. Environmental objects are thus represented as type 1/2 environmental systems, with the explicate order beyond our skins as a type 1 system, our involuntary, prereflective transformations of it a type 2 system, and the closure of their interaction a type 1/2 environmental system.

The meanings generated at the first level of the isomorphic sign are fundamentally integrated, nonreflective, symmetrical, and anonymous. These kinds of meanings are presymbolic and presignifying and therefore ineffable, and we live most of our lives within them.

At this most fundamental level of meaning there is the kind of unconsciousness that Freud described in the *Interpretation of Dreams* as having the status of the primary process in psychic life. An example of this kind of meaning is the meaning that comes with bodily sensations or from sensorimotor activity, such as the stimulation involving the senses and muscular activity. This kind of meaning represents the only kind of meaning for the infant who has not learned to speak. For the adult it represents the major kind of meaning in any field of experience. Bodily sensations and stimulations can be represented symbolically, but they themselves are not symbolic, nor can they ever attain that level of signification and expression. Pain or the internal sensations of, for instance, a stomach ulcer, or the stimulation of one's ears and eyes by environmental vibrations, or the muscular coordination needed to hit a tennis ball with a racket, can all be spoken about, as I am doing here, but as ineffable patterns of meaning these activities are directly inexpressible. These meanings, which can be felt or experienced directly, cannot be said directly but can be "shown" (in Wittgenstein's terms) by being experienced. These sensations and experiences are inexpressible, but they are not *meaningless.*

The character of this anonymous subjectivity is easily identified in infants who have yet to learn to speak. As Merleau-Ponty tells us: "During the first months of life, the child cries; he makes expressive movements; and then he begins to babble. One must consider this babbling as the ancestor of language."[7] Merleau-Ponty suggests that this babbling is extraordinarily rich and includes phonemes that do not exist in the language but that come from the child's environment. For the child, however, the babbling is not reflective symbolization or the construction of phonemes but spontaneous gestures that spring from an incarnate and anonymous subjectivity. Around the seventh month the free babbling appears to be transformed, little by little, into a voluntary effort to speak. Jacques Lacan has categorized this gradual transformation from babbling to words as the symbolic order and has given it patriarchal and oedipal significance by calling it the *name of the father*. Lacan reverses the order of the isomorphic sign by privileging the symbolic over anonymous subjectivity, for he tells us that "it is the world of words that creates the world of things."[8]

In contrast, that other French philosopher, Merleau-Ponty, indicates that it is artificial to draw an absolute division between an infant's first word and what has gone before, for the child has already been defining and referring to objects by gestures other than words. But it is also arbitrary for Lacan and other formalists to privilege signifiers over the bodily processes that produce meaning. Yet this clear-cut erasure of prereflective meaning that informed structuralist and poststructuralist writing provides the groundwork for the traditional psychological division between language and thought.

If we were to accept that "language" begins with the symbolic order, then what has gone before can be dismissed as of secondary importance. Or in the case of rational psychology, what the infant experiences before the formation of words can be viewed as occupying the realm of thought separate from language. This latter view we find in Piaget's writings on thought and language. Thought in this case is a different operation from language and represents a kind of "transitivity of relations" that produces nonverbal puzzle-solving behavior in infants, like playing a game of finding a watch under a blanket.[9]

Piaget concludes that language and thought in infants are linked in a genetic circle in which each leans on the other in an interdependent formation and reciprocal action. This account matches Merleau-Ponty's description of the relationship between the gesture of words and the anonymous gestures in infants that precede the formation of words.[10] Yet by using Piaget's description of "thought" and "language" we reproduce the unnecessary but classic division between two functions that are simply different integrated features of the isomorphic sign. An infant's cry, babble, laughter, and other expressive movements are different from the child's formation of words, but they are not so different as to demand a division by category (thought as distinct from language) or to be dismissed as irrelevant. Nor are they so different as to demand Julia Kristeva's terminology of "potential meaning functions."[11] The notion of "potential meaning" indicates that meaning begins with the symbolic, which is precisely the division that is argued against here.

The prereflective consciousness of the sensory body is different from reflective subjectivity involving higher centers in the brain, but the difference does not demand two divided and separate categories or the erasure of one in favor of the other. The differences are of gradations within an integrated discursive system.

In the intellectual traditions of Western rationalism, anonymous, prereflective consciousness is mostly masked by what is considered to be the primacy of signifiers (the second level of the isomorphic sign). Prereflective consciousness is generally eclipsed by the idealism of what is seen as the symbolic in rationalist interpretations of adult behavior. This perspective has traditionally privileged the symbolic at the expense of what Merleau-Ponty calls "incarnate subjectivity." We see this erasure of the body by the symbolic in Saussure's writing and expressed directly by Derrida: "there is nothing outside of text" [there is no outside-text; "il n'y a pas de hors-texte"].[12]

These three writers are not unique or special, for they have simply carried on the long tradition of distrusting the body or erasing it from any intellectual or philosophical discussion. In Kristeva's writing we find an attempt to come to terms with this erasure through the acknowledgment of an infant's anonymous bodily gestures that she designates by the term "semiotic chora."[13] But this designation is bracketed by her qualification of the "potential meaning" of the semiotic chora. Meanings that are prereflective are not potential, they are actual. Such terms as "potential meaning" therefore undermine the very thrust of Kristeva's intentions.

The ineffable prereflective meanings of the body are not only the first level of the isomorphic sign, they are the central and only site for the production of discourse. Monica Langer reminds us that even in discourse we do not preface our actions, our perceptions, our hearing, our use of vocal cords, throat, and facial muscles with reflections on the position of our limbs, the location of muscles, or the manner in which our eyes work. As she says, "it is perfectly obvious that for the most part we carry on our lives at a prereflective level."[14] Involuntary, prereflected actions dominate our entire body in its every motion, gesture, or expression, so that we live our multifarious experiences on an immanent, prereflective level of consciousness. (And when it comes to consciously learning a set of coordinated actions, such as playing tennis, it is always best if we try *not* to reflect on the action when it is being performed. Here we actively seek prereflective consciousness.) Yet this overwhelming, anonymous, and ineffable system of meanings is normally denied or erased in Western rationalism and specifically by structuralist and poststructuralist linguistic theory.

The Body Context

The individual body is always the context for the production of discourse and culture, and its location in space and time determines the social and political strategies embedded within discourse. For example, the social occasion in which persons find themselves will largely determine the rhetorical style and elements of expression; a football match occasions different expressions than does a job application. The emphasis on specific contexts contrasts with Saussure's sign. With this binary approach there is no context for the incarnate subjectivity, and as a consequence the effects of specific social contexts have also tended to be ignored.

The Saussurian sign is idealized by being disembodied and unlocated in space-time; so too is *langue* unlocated. As described they are independent of a physical or social context, and so the binary sign and the pair *langue* and *parole* tend to logically reinforce each other. Yet while such a reinforcement of binary pairs produces an erasure of the body context, it also produces a separation of *langue* from *parole*. The social object language is thereby divorced from individual attention, meaning, and speech. As a social medium, language is seen to exist independently of the individual and individual expressions.

This rational model of language is built on the separation of the social and the personal: of *langue* from *parole*. From an isomorphic view these binary separations are illegitimate, for the social and the individual elements of discourse are inseparably and organizationally entwined. Translated in Saussure's terms, *langue* is only made manifest through some form of *parole*, through some form of expression. Thus the social elements of language (seen as code types, such as English, French, German) are reproduced only by individuals in the series of expressive gestures that is discourse. In other words, the type of signifiers will be determined by the social, that is, the linguistic environment, but the actual (re)production of the type will always be determined by the system in general of subjectivity: by the individual in a series of individual expressions. The production of any discourse will therefore be a synthesis, a dialectical exchange involving the social type and the personal expression.

In this integrated model of discourse, the social cannot be reduced to the personal or vice versa. These two are different, yet inseparable. System type can never be determined by the individual, but the (re)production of type is always constructed by the individual. In other words, we are free to construct discourse, but only by using the alphabet, vocabulary, and syntax that result from a history of social interaction.

From this dialectical view, discourse is the synthesis of personal gesture and code types: more generally, a synthesis of system type and the system in general of subjectivity. This dialectical connection means that Wittgenstein's argument about the impossibility of a private as opposed to a public language is irrelevant. There can be no private language, as there can be no public language. Both possibilities are based on a binary opposition that is demolished by an isomorphic view of meaning, in which so-called language is always a discursive synthesis of both public and private. From this perspective language as a formal entity existing beyond the individual in some form of library or dictionary is not only irrelevant but impossible. If *langue* is situated, embodied, and human, and if we are to use *langue* as a category at all, it has to be a category of memory, not a formal, "objective" environmental entity in its own right.

Thus the potentials of *langue* are the potentials of individual histories, of learning, not some "given" formal, institutional set of objects. Language as discourse represents expressive gestures: the imitation, repetition, and the constant reassembling of signifiers by the individual. If two people use the same words and the same rules of syntax, this does not mean that language is a tool that different people can pick up and use like a hammer. Language, in the isomorphic sense, represents a series of recursive symmetries in the form of mimetic repetitions of alphabet, words, syntactic rules, narrative conventions, and generic considerations that are combined by individuals differently. These mimetic repetitions represent social symmetries: habits and patterns of conformities necessary for any exchange to take place.

What is interesting and surprising about discourse is the degree and amount of social conformity (of mimetic repetition) it involves. This recursive symmetry is seen in the habits of expression that involve the use of the same alphabet, rules of syntax, sentence "fillers," and often vocabulary. And yet different individuals will combine these habits differently (nonsymmetrically) in the form of different signifieds, concepts, or images and different rhetorical styles and strategies. It is this dialectical interaction of the symmetrical and the nonsymmetrical elements of discourse and meaning that make this expressive gesture so individual and yet so social.

Incarnate subjectivity of the body automatically creates a second form of context for discourse. As the body is always situated physically, environmentally, and socially in some particular time and space, these differential factors provide a social context to any expression. Yet the meanings provided by the social context will be ones that are overwhelmingly prereflective. How is this? These meanings will be made up of mostly the visual and auditory effects of environmental conditions as well as other images and signs. But while images and signs are representational forms, they will not normally be treated reflectively while awareness is focused on the production of a particular discourse. Instead, images and signs, along with the environmental conditions, will be mimetically reconstructed by the individuals involved in the discourse. Such mimetic reconstructions represent prereflective meaning.

Any discourse will therefore always be contextualized, that is, given a specific meaning, by location and social setting. The relationship between social context (sometimes called a genre) and expression will also always be a dialectical interweaving in which the smaller "whorl" of expressed meaning will be situated within a larger "whorl" of the prereflective meanings of the environmental and social context.

Conclusion

In this chapter we began to discuss the integration of an isomorphic sign. This structure represents the three levels of subjectivity involved in the production of discourse. As an isomorphic approach always has a focus on individual subjectivity as a constituent of the context of cosmic consciousness, discourse is seen here as the manifestation of subjectivity in cultural forms. Discourse thus represents the third "layer" of consciousness that has emerged from the context of the system in general of subjectivity, which in turn has emerged from the context of the symmetries of cosmic consciousness. Discourse, therefore, is not a formal separate entity like the fiction "language," but a recursive and integrated set of representational and rhetorical processes that involve each of the three levels of the isomorphic sign.

Chapter 11

Signifiers and Signifieds

In this chapter we discuss the implications of signifiers and signifieds within an isomorphic sign structure and compare them with a Saussurian view. Traditionally, Saussure's sign represents "language," but it is also a model for rationalist interpretation. His sign has two "vertical" levels: sound images and concepts. He renamed these as signifier and signified. This interesting implication of psychological processes (seen in the terms concept and sound image) with linguistics was never followed through to its conclusion by Saussure or his followers, most probably because the formalism of his linguistics overwhelms the psychological. It is this connection of psychology to linguistics through semantics that is the focus of our present discussion.

Interpretative Machinery

An isomorphic view of Saussure's sign resites this structure within the parameters of the field of meaning produced by subjectivity. One of the purposes of this relocation is to reevaluate traditional theories of language. This means that we are not interested here in traditional cognitive-science questions that seek to trace language processes in the brain. We now have relatively detailed maps of the brain when it is involved in language activity, for example when we hear, see, and speak. Interestingly also, there appear to be three interacting neural regions that can roughly equate with the three levels of the isomorphic sign.[1] Yet while this neurological material is interesting, it does not provide a workable description of language or reflect on the implicit meaning or the asynordinate structure of language.

These questions are essential if cognitive science is to move beyond its instrumental traditions. For example, the questions about language affect our view of the connections between subjectivity and the nervous system. For instance, if we accept that meaning is produced by the combined physiological mechanisms of excitation and inhibition, then signifiers should be formed by those centers of the brain where inhibition is dominant. This should be the case because signifiers represent the level of the sign at which differences are dominant. As signifiers are characterized by abstract and formal differences, in contrast to the distinctions of manual manipulation, the brain centers involved in the production of signifiers will be distinct from other specialized sensorimotor centers.

To question the structure of language rather than trying to map this instrumental term onto the brain is to endeavor to change some practices of traditional cognitive science. If our model of manifest subjectivity has a recurring structure sufficiently complex to appear isomorphic with neurological patterns, then the process of "fitting" should become easier. The problem in the past has been that we are saddled with an inadequate theory of subjectivity and a Platonic view of language, both of which have been detrimental to the integration of neural and semantic processes. With these comments in mind, we begin to look at the microstructure of signifiers and signifieds.

Signifiers

What are the isomorphic characteristics of signifiers? There are several that vary from the features Saussure attributed to signifiers. One general consequence of the reason that frames Saussure's linguistics is that a structuralist interpretation is profoundly antihumanist. The erasure of the category of subjectivity from Saussure's sign, an erasure also central to poststructuralist theory, leads directly and inevitably to formal and antihumanist interpretations celebrated by such writers as Stanley Fish, who has produced a familiar formalization: "there is no subjectivist element of reading because the observer is never individual in the sense of unique or private, but is always the product of the categories of understanding that are his by virtue of his membership in a community of interpretation."[2] An isomorphic view of signifiers challenges this antihumanist tradition.

Isomorphic signifiers occupy the second level of the sign and not the first, as they do in Saussure's sign. The ramifications of this resiting are profound in terms of paradigmatic frameworks. Signifiers are viewed here as expressible linguistic units that emerge from the prereflective subjectivity of the body. This humanist resiting means that all signifiers are both psychological and linguistic in their construction and their use. Therefore, signifiers such as sound image or written letters of the alphabet are expressible, asynordinate units involving both the sensory and motor processes of perception and verbalization as well as memory.

It should be noted that the isomorphic connection of subjectivity and linguistics is not a parallelism, but a synthesis. The effect of this synthesis, in a more general sense, means that the structure of discourse and interpretation is also the structure of manifest consciousness, and vice versa. The following sections describe some of the features of this isomorphic interpretation of signifiers.

The Integration of Signifiers

The paradigm of reason is primarily oriented to differences, and traditionally the differences of signifiers have been significant elements in the formalism this perspective generates. If study in any field has a focus and an orientation on the signifiers of that field, this study will involuntarily reproduce a rigid formalism along with a rationalist perspective. This is the case with a Platonic view of mathematics, with most linguistics, with much statistical psychology, and with rigid bureaucratic public administration, and so on. Signifiers, above all else, are differential marks within conscious awareness, and the most prominent foregrounded feature of these marks is their difference from other signifiers and from the background of sensory and felt experience. The meaning of difference, it will be recalled, is the dominant structural feature of conscious awareness, as well as the dominant feature of signifiers. The symmetry of location and difference makes signifiers one of the constituent features of conscious awareness.

For the discipline of linguistics and structuralism, in particular, we find an explicit celebration of reason's bias in Saussure's linguistic model: "in language there are only differences."[3] Signifiers occupy a primary place in the sign structure, and so a Saussurian view is concerned with the distinctions of signifiers. As Jonathan Culler points out in his introduction to the *Course in General Linguistics,* "one may write *t* in numerous ways, so long as one preserves its differential value. There is no positive substance that defines it; the only requirement is that one keep it distinct from other letters with which it might be confused."[4]

From an isomorphic view, this statement is both inadequate and incorrect. It is incorrect insofar as the signifier *t,* like all other signifiers, does have a "positive substance," and it does have an internal structure in that it is an asynordinate system in its own right. Its construction in subjectivity, by the nervous system, represents a series of processes that can be seen as a system. As all systems have the universal and asynordinate structure of symmetry, nonsymmetry, and asymmetry, so also does the letter *t.* For example, the system *t* appears in conscious awareness in a definite form, on a background from which it is distinct, and within a larger system (which is linguistic) of which it is a constituent part. The unconscious relations of symmetry recursively construct its location within the linguistic system (the constituent-context relationship) as well as constructing it (through the neurological processes of excitation and inhibition) within the processes of perception. The symmetry of the projected percept with remembered image represents the major and primary mimetic representational relationship of *t*. In the rationalism of Culler and Saussure these primary symmetrical meanings are simply erased along with the perceiving subjectivity that constructs and expresses *t*.

Saussure's rationalist linguistics indicates that both the linguistic unit and the value of difference are unanalyzable primary elements of language. For instance, a signifier such as the grapheme *t* is seen by Saussure and Culler to have no internal structure beyond the value of difference, which is also unanalyzable. The unit and the value of difference tend therefore to be indistinguishable from one another, for what distinguishes a unit from other units is their differences. This synthesis of unit and the value of difference is summed up ambiguously by Saussure: "value is made up solely of relations and differences with respect to the other terms of language."[5]

From an isomorphic view, Saussure's comments have the general lines of argument right, but not the detail. The linguistic unit is not a synthesis of only difference, it is the synthesis of a series of symmetries and differences that together occasion the unity of meaning and unit, and in addition, the other units in the linguistic system. But that is not all: the dialectical intertwining of symmetry and difference does not lead to an erasure of difference but to a relative balancing of it. A linguistic unit therefore has distinction of form, set as it is on a background context, and, in addition, a distinction as a constituent within a larger linguistic context. The meaning of a linguistic unit like *t* is therefore not only differential but has a balance of symmetries and nonsymmetries in figure/background, constituent/context proportions.

The isomorphic sign is not just an interpretative model for linguistics, it is also a more general model for viewing any behavior. A conscious awareness that is oriented to surface and differential features of signifiers will thus tend to believe in differences, divisions, separations, and the conventions and rules that give emphasis to these differences. This perspective is found in such diverse places as the truth tables of symbolic logic, the social exclusion of economic classes, the disciplinary rules in child rearing, the differential calculus, and, in extreme and gross forms, in racist behavior.

In the diversity of these examples, the term *signifier* is used in a broad semiotic, rather than a narrow linguistic, sense. In this broad semiotic sense a focus on signifiers represents both a trust in their surface appearances and a valuation of their differences. In other words, this focus tends to produce an "instrumentalization" of complex processes and a trusting belief in explicit differences.

Such contentment with the instrumental is common to a variety of contemporary behavior. Consumerism and technocratic and technological practices embody a tendency to overvalue differential appearances and the instrumental, but in various ways so do the practices of disciplinarians, deconstructionists, and social elitists, as well as racists and many other modern agents.

Signifiers are thus both the strength and the weakness of conscious awareness. A conscious awareness that is dominated by the irrationalism of symmetricized meaning (too much affect) will need more differential signifiers to transform and move the affect. Yet a conscious awareness that relies too heavily on signifiers will tend toward a superrationalist, fragmented, and alienated subjectivity. Both linguistic and semiotic signifiers therefore need to have a balanced treatment, so that their differential surface appearances are embedded within felt, implicit meaning (the heart).

The Social and the Personal

A second difference between an isomorphic and rationalist interpretation of signifiers centers around the relationship of the social and the personal. From a rationalist perspective this relationship is always a problematic one, whereas from an isomorphic view it represents just another dialectical synthesis. The difficulties of a rationalist approach to this synthesis are apparent in Saussure, who states that "language is a social fact," but also that "everything in language is basically psychological."[6] This relationship of the social and the personal is, for Saussure, inevitable: "Speech has both an individual and a social side, and we cannot conceive of one without the other."[7] But what in a signifier is social and what personal? Saussure provides an illustration to draw out their differences: "When we hear people speaking a language we do not know we perceive the sounds but remain outside the social fact because we do not understand them."[8]

From these remarks we could assume that the social aspect of signifiers has something to do with learning and understanding a particular language like French, while the individual side is related to the execution of a speech act. In a further attempt at defining the social aspects of language, Saussure identifies words with the social: "If we could embrace the sum of word-images stored in the minds of all individuals, we could identify the social bond that constitutes language."[9] Should this have been "sound images" instead of "word-images"? This is a slippage we will return to shortly.

Saussure tell us that sound images are auditory impressions and that "auditory impressions exist unconsciously before phonological units are studied; our ear tells us what *b, t,* etc. are."[10] In other words, according to Saussure, our ear divides and delimits a continuous series of sounds into a linear system of auditory impressions called sound images. However, this materialistic interpretation of sound images is subverted to a large extent by Saussure's later description of the method by which language is delimited. "Delimitation" refers to the way language is made up of units. Saussure does not relate the units of language (sound images) to his earlier statement concerning the auditory limits that the ear constructs. Instead, he ascribes the determining influence to concepts, signifieds: "In an accurate delimitation, the division along the chain of sound images (a, b, c) will correspond to the division along the chain of concepts (a′, b′, c′)."[11]

The confusion over the delimitation of language arises because his earlier, materialistic interpretation conflicts with his later, idealistic descriptions. This confusion carries over into the nature of the relationship between signifiers (sound images) and signifieds (concepts), to be discussed shortly. From an isomorphic perspective Saussure's first interpretation is closer to the mark than his second, Platonic one. The conceptual determinism of his second interpretation reverses the whole order of the isomorphic sign by placing the determining value of language with the last level: concepts. On this point Roman Jakobson has disagreed with the inherent Platonism in Saussure's second interpretation of how language is split into units. Following from Saussure's first description, Jakobson refers to the character of sound and how this overlapping and undifferentiated process is transformed into successive sound images:

> From a strictly articulatory point of view there is no *succession* of sounds. Instead of following one another the sounds overlap; a sound which is acoustically perceived as coming after another one can be articulated simultaneously with the latter or even before it.[12]

These processes of unconscious, undifferentiated, and sometimes overlapping vibrations that affect the "normal" ear can now be demonstrated and measured by such devices as the spectrograph (which is sometimes used in the forensics of voice identification). The transformation that occurs when simultaneous and overlapping vibrations beyond conscious awareness are changed into differentiated, successive sound images that are formed in conscious awareness is an interesting and extremely rapid set of integrated processes performed isomorphically by the nervous system. I will return to these processes of transformation shortly, but in the meantime the question of the relationship of the social and the individual remain.

What is it about these personal, perceptual processes that is social? Neither Saussure's several interpretations nor Jakobson's auditory description sheds much light on the question. In the case of Saussure, who addresses the question of the social directly, I suggest he is unable to define the social aspect of signifiers because the rationalism that frames his view precludes an integration of apparent differences. This rationalist tendency is also at work in much of Jakobson's approach to meaning. Needed here is an isomorphic perspective that can accommodate an integration of two apparently different features, the social and the personal.

From an isomorphic perspective the social and the individual aspects of signifiers (sound images) are readily integrated together and yet distinct. The social aspect of signifiers represents the system *type* of signifiers, whereas the personal aspect represents the *system in general* of meaning that mimetically reconstructs that type of signifier in the conscious awareness of the receiver. Types of linguistic signifiers differ from one another, as French and English differ from each other. Yet the organization of the system in general, a system that constructs and reconstructs these types of signifiers in discourse, remains invariant for all "healthy" individuals. In other words, the organization of the constructing system that accounts for discourse and interpretation remains invariant during the transformations involved in expressing the social type. The social and the personal are in this manner inseparable partners in the formation, construction, and expression of every signifier.

Language and the World

A further divergence of views on signifiers results from different interpretations of the relationship language has to the world. Structuralists assume that the physical world has no structure and that it is only given structure by language. In contrast, an isomorphic view accepts that the world has structure and that its structure represents the element of type in any system. Isomorphically, the relationship discourse has to the world initially comes down to the question of how signifiers are constructed in subjectivity.

Signifiers, whether auditory or visual (whether phonemes or graphemes), are constructed in subjectivity in similar ways. The transformation of sound into sound images follows a path similar to the transformations that occur in vision when we construct a visual linguistic unit such as the letter *t*. It has been shown in experiments that when subjects are exposed to an array of letters of the alphabet with exposure times of 50 milliseconds, they report that letters appear present and legible when the stimulus had actually been turned off for 150 milliseconds.[13] From the point of view of the experimenter such performance is based on "memory," while from the subjective experience of the task it is a "perceptual" one. The visual persistence of these images lasts about one second, depending upon such variables as intensity, exposure time, and postexposure illumination.[14] In these experiments the presence of memory in perception is central to the manner in which signifiers are constructed in consciousness.

These psychological experiments suggest that memory and perception are integrated and operate in a similar manner for both sound images, which have auditory pathways, and the visual images of the letters of the alphabet. These experiments imply that in terms of simply constructing a letter of the alphabet (of identifying the letter *t*, for instance) the visual processes involve a transformation that symmetrically links perceptual and memory processes. This linking can be as fast as 50 milliseconds or take up to one second. Similar complex auditory processes are most probably involved with the identification of sound images.

We would hypothesize that in terms of the structure of meaning, the transformation from sound to sound image could be brought about by the successive closures imposed on the stream of overlapping and simultaneous vibrations (sounds). These closures would produce "delimitation," as in a series of differentiated, successive, conscious sound images. The delimiting of undifferentiated sound does not happen in the ear, as Saussure originally suggested. It most probably involves a series of pathways that begins with the combination of excitation and inhibition in the ear involving the auditory pathways and reaches symmetrical completion in the Wernicke's area of the brain, where traces of sound images maybe stored and where the identity of new sound images is constructed.[15]

These processes of construction (of identity) can give rise to several interpretative possibilities. The first is the structuralist position that asserts that language structures the world. Isomorphically, such a view is incorrect. The nervous system does not arbitrarily construct and impose a sound image on the incoming, unstructured vibrations, as would be the case if the world were structureless. In other words, the sound images that are identified in the Wernicke's area of the brain are not randomly brought into consciousness and arbitrarily imposed onto the incoming undifferentiated sound. If this happened it would be similar to stimulating the Wernicke's area of the brain with an electrical probe. In such cases we would hear successive sound images that had no symmetry with sensory stimulation. We could therefore "hear voices" that were not related to the physical vibrations in the air and recorded by the ear. Such a situation is akin to having an hallucination, which is very different from normal perception.

Yet a further reading of these processes may be the opposite to the solipsistic view above, and that would be an objectivist interpretation. Such a view would say that signifiers (like all other objects) are entirely physical objects. Such a view is equally incorrect. As Jakobson points out, sound images are not carried on the air waves, nor are they heard by the ear. A more subtle version of this approach would view the undifferentiated vibration, which stimulates the ear, as determining the nature of the sound image. This (soft objectivist) interpretation is also inadequate, for if this were the case, then language (and culture) would be entirely opaque and innocent, and physical "reality would be absolutely determinate in all cases.

As anyone vaguely familiar with linguistics will appreciate, this cannot be the situation, which means that the structure of the physical vibrations will not absolutely and fully determine how sound images are constructed or in what order they succeed one another. On this point Saussure's comments regarding the understanding of a foreign language are apposite. If we hear a foreign language, we hear unfamiliar sounds and therefore are unable to take the next necessary step of constructing sound images that match up with the structure of the incoming vibrations. If the objectivist position were correct, there would be no such thing as a foreign language, for we would automatically understand any and all language.

The binarism that subtends both the solipsistic and the objectivist viewpoints is inadequate to account for the auditory transformations that occur in the complex, dialectical processes of hearing sound images. From an isomorphic perspective, sound images are constructed through the semantic processes of closure. Closure results from an almost total symmetry, involving the coupling together of at least two complex systems. The closure that constructs units of sound (sound images) in the process of hearing a language we know is a closure that involves a mimetic projection onto the environment. Its mimetic features are those that match up known sound images (stored in memory) with the structure (the element of type) of incoming sounds. The stored images become mimetic of the sensory effects of physical vibrations through the process of a symmetrical coupling. In this respect, both elements (the stored and the presented) are active agents in determining the final synthesis, and then closure, of sound images.

This is a process of matching up auditory images in memory with the structure of incoming vibrations. This "sorting mechanism" (which functions on the basis of the interconnection of symmetry and nonsymmetry) happens "instantaneously" (probably at about 50 ms), so that there is no conscious interruption to "hearing." Insofar as the element of projection is concerned, the symmetry of these two systems (the stored and the presented) means that the sound image appears in conscious awareness as coming from a direction in the environment. Our ability to determine the source of any sound will depend on how mimetic our projections are. As in the processes of vision, we thereby project outward what has been a constructed synthesis inside our nervous system. Again like vision, the mimetic projection of a sound image lacks definite spatial coordinates: in this sense it is spaceless. Yet this spacelessness is not experienced by awareness during perception; rather, there is an experience of certainty in the construction of the closures of perception.

A further aspect of this particular closure of sound with image is that it involves two types of systems: one learned, and one presented by the environment. The closure itself, however, is due to the recursive mechanisms of the symmetrical organization of the overall system in general of meaning. While the concept of type of system may seem applicable to sound images that are learned and then stored in memory, the term *type* of system may appear dubious when it is related to vibrations emanating from the environment. Can physical vibrations be described as a type of system, that is, having some kind of structure? In the wider sense, we can ask the question: is the world structured, or do we construct its structure?

The reciprocal and isomorphic relationship between subjectivity and the world means that the world has the same asynordinate structure as consciousness. Therefore, the reality of speech is constructed by a mimetically determined relationship of remembered sound images and incoming vibration. The implication of this isomorphism is that the physical world does have a structure that we reproduce and involuntarily project, at a prereflective level of perception, onto the environment. This proposition contrasts with a structuralist and poststructuralist perspective that always foregrounds the differences between these two seemingly separate domains of language and the world.

We find this assumed separation in Jakobson's comments about how sounds consciously perceived as successive can yet be simultaneous with one another when they are measured as physical vibrations. The implication of these comments is that the successive order of sound images is a separate domain from the physical vibrations, which seem structureless and orderless. Jakobson's view here coincides with an overall structuralist and poststructuralist position on this question. This view is expressed by the statement that language gives form and structure to vague unstructured life. Thus Saussure is able to say that "without language, thought is a vague, uncharted nebula. There are no preexisting ideas, and nothing is distinct before the appearance of language."[16] In a similar vein Catherine Belsey tell us that "the world, which without signification would be experienced as a continuum, is divided up by language into entities."[17]

Similar views were expressed by the linguist Benjamin Lee Whorf. Whorf refined Edward Sapir's thesis of linguistic relativity, which then became well known as the Sapir/Whorf hypothesis of linguistic relativity. It proposes that the native tongue of an individual determines how the world is perceived and how individuals think and feel about it. Each language constructs its own "habits of thought," which are made up of "the microcosm that each man carries about within himself, by which he measures and understands what he can of the macrocosm."[18]

While each of these writers approaches the subject slightly differently, they nevertheless hold the common view that language structures the structureless nebula of the world. From an isomorphic perspective the binarism that informs their views is unacceptable. The argument that we have to choose between the physical world as an active and determining agent or language as an active and determining agent does not fit with an isomorphic paradigm. Insofar as auditory vibrations are concerned, there is a structure type in these vibrations, and this type has a symmetry, at given points, with the structure of the internally stored sound image. The structure of vibrations will not be identical, in all conditions, with sound images. But while the two structures are not identical in every respect, they must possess sufficient symmetrical relations to ensure a permanent closure of sound with image; otherwise there would be no such thing as sound images.

This description represents the usual conditions of hearing a known language. It does not represent the unusual conditions of mishearing a known language. In this situation the closure of image with vibration does not represent a case of lack of symmetry but of insufficient symmetry (too much nonsymmetry) to warrant a closure. Such inadequate closures can be due to deafness, unfamiliarity with dialect, or indistinct sounds. Like "jumping to conclusions," a mishearing is not a negation of the symmetry of sound and image but an example of insufficient symmetry to warrant the closure produced. Mishearing is therefore an aberration of the "normal" symmetry of types: an aberration of a system geared to producing closures at a frequency rate of approximately one every 50 milliseconds.

I have used the term *transformation* to describe the change that occurs from overlapping and simultaneous sound into successive sound images. The concept of a transformation indicates that changes take place within the invariances of the auditory processes. A transformation does not imply a split, a separation, or a series of differences that divides two domains. In other words, it does not imply *only* differences. Nor does it imply *only* symmetry. A condition of only symmetry would lead to a simple linear extension of sound into sound images without transformation. It would be a replication, as indicated by an objectivist approach.

The mimetics of sound and image is therefore not a *pure* symmetry divorced from the rigors of nonsymmetry. A transformation implies the rather active interaction of two ingredients, symmetry and nonsymmetry (excitation and inhibition). In their active, competitive integration, these two semantic elements lead recursively to two possible outcomes: a correct hearing, or a mishearing of sound images. In either case, however, the closure of system types—of vibrations and sound images—is not a closure brought about by a simple symmetry without qualification. Nor is it a case of the negation of symmetry resulting in different domains. Rather, it is an isomorphic synthesis in which the relation of symmetry dominates nonsymmetry.

To hear a sound image correctly, therefore, is to mimetically transform vibrations into sound images constructed roughly symmetrical with the structure of vibrations. This transformation is therefore not a simple linear replication. In our discussion of perception in chapter 8, the same isomorphic conclusions were made in regard to visual images. It will be remembered that the mimetic projection of percept results not from linear replication but a transformation of environmental structural types.

What is a Signifier?

Yet another divergence of views relates to the reason's difficulty of defining a signifier. This difficulty is also related to those Platonic tendencies that unconsciously slide the focus upward toward higher and higher abstractions. The difficulty of description comes about because reason foregrounds distinctions, and signifiers appear distinct, yet the condition of distinctness seems to be beyond analysis. The effect of this is that all linguistic units are assumed to be signifiers because all linguistic units are distinct from each other. This generalized classification of all linguistic units as signifiers is incorrect.

To define the limits of isomorphic signifiers it is necessary to begin with some of the psychological effects of a focus on signifiers. An exclusive orientation to signifiers invariably produces an unconscious and Platonic transcendence of signifiers into concepts and images. A slippage like this, from signifier to signified, is always the author's desired effect in fiction and narrative. Fiction is designed to weave illusions and to make readers forget that they are reading letters, words, and sentences. This managed slippage into the play of signifieds is to be expected and possibly desired in fiction. But what is an appropriate, merely temporary blindness in fiction is in nonfiction an unfortunate symmetricized slippage.

Instances of transcendental slippage are common even in rationalist and logical mathematicians. Davis and Hersh remark that "the Platonist conception of mathematics (held by nearly all mathematicians nearly all the time) attributes to mathematics certain properties and features usually attributed only to God."[19] For such mathematicians, mathematics is independent of the world; should the world some day come to an end, mathematics, like God, would continue. Both God and mathematics are instances of self-existing, self-knowing, and self-justifying systems. According to Davis and Hersh, "God is a mathematician" is the slightly egocentric way physicists "answer the unanswerable question of why mathematics is such an effective tool in theoretical physics."[20] Thus if the world is mathematical, and it appears that way to some, then if there is a God, he must be a mathematician.

The question that concerns us here is not the theological argument about the existence of God, but the process of transcendent signifiers, whether they are numbers or linguistic units. The Platonic view of mathematics conceives of $2 + 2 = 4$ in a formal and Saussurian manner. Such a perception erases the subjectivity that expresses this text, and the effect of doing so is that the "horizontal" series of closures that inhere in this statement is condensed and rarefied into a transcendental truth. Such a truth has the features some theologians have attributed to God. This conclusion derives from rationalist interpretive strategies. These strategies hold that in any reading or language, subjectivity is erased. Such a strategy will, of necessity, focus exclusively on signifiers and thereby slide quickly and contradictorily into replacing those signifiers with "free-floating" signifieds. This slippage may happen because this focus is ungrounded by context (subjectivity), and so an ungrounded focus will always tend to slip from feature to feature.

Without the context of subjectivity, this slippage seems the inevitable result of a rationalist focus. Given that Saussure erased subjectivity from his sign structure, we can expect evidence of this slippage within his analysis of language. This evidence is apparent with his difficulty in defining a signifier.

Saussure described language as having two dimensions: a "vertical" axis and a "horizontal" axis. These axes are ubiquitously coordinated. The horizontal he called the "axis of succession," and this gives rise to the succession of signifying units within the linear dimension of expression, for example, in the four letters in the term *w-o-r-d*. The vertical is called the "axis of simultaneity," which, by its nature, excludes the differential and extended elements of time. An isomorphic perspective generally agrees with this description of the two axes but would rename them in terms of the relations of meaning, as the vertical axis of symmetry (simultaneity) and the horizontal axis of nonsymmetry (succession). From an isomorphic perspective these two axes are ubiquitously synthesized in discourse.

While Saussure tells us that these two axes are coordinated, he is also inclined to highlight their differences by referring to them in terms of "synchronic" and "diachronic" linguistics. Such descriptions tend to imply the separation of the two axes of language. The question arises, however: If a signifier is linear, having a horizontal dimension (and this is generally accepted by linguists to be the case), what is it about the signifier that accounts for its axis of simultaneity? Not surprisingly, Saussure was unable to account for the signifier's axis of simultaneity. From an isomorphic perspective, this is its symmetry. As symmetry becomes apparent only when the category of subjectivity forms part of our interpretative model, and as the concept of subjectivity is missing from Saussure's account of language, it is no wonder he had difficulty explaining a signifier's simultaneity, that is, its symmetry.

This critical question of a signifier's simultaneity has been raised by Jakobson in relation to the possible atemporality of language generally.[21] In addition to the problem of atemporality (which I refer to shortly), I suggest that the signifier's simultaneity also raises the problem of what constitutes a signifier and how a signifier is different from a signified. These are questions related to the problem of slippage, and how the continual transcendence of signifiers by signifieds is brought about by Saussure's rationalist and linear character of language.

To discuss this problem of slippage in Saussure's linguistics, we begin with the "linear character of language." Saussure taught that "the linear character of language precludes the possibility of pronouncing two elements simultaneously."[22] He also said that the elements of language "are arranged in sequence on the chain of speaking";[23] combinations of elements are called "syntagms." Syntagms are always composed of two or more consecutive units, for instance, the French *re-lire* "reread." From these comments and examples, we see that Saussure's language is linear.

In light of this linear character of language, we ask, "What constitutes a signifier?" Is a signifier a single unit of sound or vision, a sound image and/or a visual image? Or is a signifier a written or spoken word, such as "red"? Or is a signifier a syntagm, a word that combines two or more consecutive units? Or is it a sentence? Now Saussure specifically said that he would replace the term *sound image* with signifier.[24] If it is the case that only sound images and their visual equivalent, graphemes, are signifiers, then the discussion of the linearity of language is a discussion that hides within its terms the question of what constitutes a signifier and how this category differs from a signified.

From an isomorphic view, signifiers are what Saussure specifically describes them to be: they are either the single visual images, graphemes (the single linguistic units we find on keyboards), or individual sound images, phonemes. If this is the case, words, whether one syllable or more, are not signifiers; neither are syntagms, phrases, or sentences. Each of these linguistic units represents successive combinations of single signifiers, and as such they represent signifieds. In addition, each of these combination units (words for instance) is constructed, not at the level of perception (at the second level of the isomorphic sign), but at the third level, by way of conceptualization. The transformation that occurs between signifier and signified is therefore a transformation from perception to conception.

To appreciate the transformation between letters of the alphabet and the words they form, it is necessary to attend to the specific features of signifiers and how they differ from signifieds. For instance, in English the letters *e-d-r,* in that order, do not "make a word"; they can only "make a word" when the order is changed to *r-e-d.* Yet what has changed? The letters of the alphabet, the signifiers, are the same in any order, and they remain more or less the same whatever the combinational order. But the semantic processes involved in writing letters of the alphabet and making a word are transformationally different; that is to say, the transformation that occurs in the construction of a letter is one that entails a necessary mimetic symmetry. The transformation that occurs when letters are combined together to make a word is one that lacks mimetic symmetry. For example, the connection between the word *red* and a color is related to community and linguistic convention (an arbitrary connection), not to the mimetic transformation of a physical structure. So there is no necessary connection here. (This point could be made adequately by turning to Saussure's argument concerning the arbitrary nature of the sign.)

There is, therefore, a contingent closure within the transformation that occurs when we "make a word." Through the processes of learning, this closure couples together a particular combination of letters *(red),* which is a type of linguistic system with a system type we have experienced when prereflectively sensing a particular color. This closure of linguistic type with sensory type is not a necessary closure, but a contingent or arbitrary closure. It is contingent on the repetitive nature of learning and the linguistic habits of a particular community. Unlike the necessary closure of the mimetic reconstruction involving letters of the alphabet or objects of the environment, this contingent closure is due to a symmetry of occasion and coincidence and can be changed without damaging the system in general of meaning or the structure of perception. Therefore, the transformational difference between constructing a letter of the alphabet and constructing a word is that the first is structured by a necessary closure and the second is structured by a contingent closure. The first is therefore a necessary truth, the second a contingent truth. The first is thus a signifier, the second a signified.

If letters of the alphabet and phonemes are signifiers, and words and phrases and sentences are signifieds, there is no indication of this category difference in Saussure's writing. Saussure did not recognize the transformations that takes place when signifiers are transformed into signifieds. In addition, as he tended to erase subjectivity from his linguistics, he did not stress the perceptual nature of signifiers in contrast to the conceptual character of signifieds.

Saussure's discussion of the linear character of language implies that there is no transformation from letter to word, only a horizontal extension. Therefore all linguistic units, whether letter, syllable, word, or sentence, appear as differing only on the basis of their length. All linguistic units, it seems, are linear signifiers. If this is the case, the slippage from signifier to signified that we witnessed in mathematical and philosophic writings is also evident here. Saussure's subject is the structure of language, yet he continually and unconsciously slips from referring to signifiers (graphemes and phonemes) into referring to signifieds (words, syntagms, and sentences) as if they were signifiers.

As stated, the cause of this slippage in these differing fields is a contentious adherence to the paradigm of reason that offers the most direct path into Platonic states. In linguistics this slippage goes relatively unnoticed because of the general adherence to reason and the resultant confusion between a signifier and a signified. Yet while this tendency for elevation is ignored in linguistics, its acceptance as theory reinforces the blindness of instrumental readings that are part of all reasoned perspectives.

The Space/Time of Language

Jakobson's discussion of the simultaneous features of language raised the prospect that the value of time is itself a linguistic value: "time with respect to its role in language proves to be a constitutive value of the latter, that is, to be a linguistic value."[25] Jakobson raised this possibility in his argument about a single-featured auditory entity called a "mora." According to Jakobson, a mora is a pointlike feature irreducible on the axis of succession (because it has no combination of features) and also irreducible on the axis of simultaneity (also because it has no subfeatures). As this description places this auditory feature outside of time (time being a differential extended matrix on the axis of succession), Jakobson concludes that "the phoneme in itself, and more generally the linguistic sign and language as a whole, are atemporal."[26]

From an isomorphic perspective I would go further and say that all language and discourse are not only atemporal but also aspatial. Without using

Jakobson's reasoned approach, we need simply ask where in space and time the letter *a* is located, or where in space and time the phoneme "a" is located. For that matter, we can also ask where in space and time discourse is located. Is it "out there" on the page (in the air for a phoneme), or is it "in here" in my head? Each of these binary possibilities represents a particular rationalist approach to analysis. The objectivist will claim that the alphabet letter *a* is out there on the page, while the solipsist will locate the letter *a* inside my head.

From an isomorphic perspective, the very transformative processes involved in the closure of vibration and image to produce the identity of a signifier mean that the explicit differential dimensions of space-time are excluded from these prereflective processes. The same conclusions can be reached in relation to the symmetrical transformations that change signifiers into signifieds. For example, the arbitrary symmetry of the linguistic system *red* and the experience of color is a contingent closure that nevertheless excludes the nonsymmetrical of conscious awareness and, therefore, of space and time. The symmetry that dominates each of these closures excludes the differential of time and space by its very isomorphic nature.

On the other hand, time and space do not stand "objectively" apart from meaning, for nothing exists outside of meaning. Time and space are asymmetrical, explicit, differential, and extended systems and therefore logically operate within subjectivity at a level of conscious awareness.[27] As was noted in chapter 2, space and time are asynordinate conditions, and their structures are mimetically reconstructed within subjectivity at the level of conscious awareness. What happens in the processes prior to conscious awareness will all be implicit nonspatial and atemporal (and unconscious), and therefore the asynordinate conditions of normal space and time are not operating. As the basic closures of discourse (the mimetic closures of signifiers and the arbitrary closures of signifieds) all take place outside conscious awareness, a great deal of discourse happens outside the conditions of space and time.[28] Discourse thus has meaning and resonance but, in a large degree, is space-less and time-less.

Thus, Jakobson can legitimately say that space and time have "a linguistic value." In addition, if the dominant feature of all signifiers and signifieds is their underlying symmetrical structure and symmetry is unconscious and implicit meaning, then Jakobson is again correct when he says that language as a whole is atemporal.

But we can go further by saying that the semantic processes that construct discourse, subjectivity, and cosmic consciousness are atemporal and nonspatial, for the foundation principle of each of these structures is one of symmetry; and as symmetries that are untransformed always preclude nonsymmetries, the differences necessary for the asynordinate conditions of space and time will also be missing. In this sense space and time as asynordinate structures of meaning are missing from much of cosmic consciousness as well as from large areas of subjectivity and also from much of discourse. Our belief that these dimensions are fundamental to the conditions of the universe is a result of our comfort with our materialist identification patterns and with the mechanistic sight lines of the paradigm of reason.

Signifieds

I come now to the third level of the isomorphic sign, which is the same as Saussure's second; it represents the conceptual level of discourse. What are the isomorphic characteristics of signifieds, and how do they differ from signifiers? In order to discuss the characteristics of signifieds we need to study their "vertical" relationship with signifiers.

In linguistics the differences between signifier and signified have traditionally been framed in the Saussurian terms of the arbitrary versus the necessary (or the motivated). Saussure implied that the two elements of the sign are necessarily linked, while stating explicitly that they have an arbitrary connection: "The bond between the signifier and the signified is arbitrary."[29] This contradiction between the necessary and the arbitrary is perhaps accentuated with a binary approach, so that we become committed to either one in favor of the other. This has been the usual result of any discussion on this question. Jacques Lacan, for example, provides an extreme commitment to the arbitrary and writes that the signifier and signified are "distinct orders separated initially by a barrier resisting signification."[30] On the other hand Benveniste attempts to square the arbitrary with the necessary but is able to do so only by shifting the argument into signification. "What Saussure demonstrated remains true, but true of the signification, not the sign."[31]

From an isomorphic perspective, the relationship between signifier and signified cannot be explained by rationalist means. It cannot be because there must be an explicit description of the integration of their differences, as implied in Saussure's notion of a motivated sign. Such a description is only possible within an isomorphic paradigm.

The following represents one such isomorphic description. Signifiers are linked to signifieds both necessarily and arbitrarily. The necessary link relates to the system in general of subjectivity; for instance, to "make a word" one has to use letters of the alphabet, and, in a more general sense, to make a concept one uses signifiers. It is impossible to do otherwise. These transformations (signifiers into signified) are performed by a healthy integrated human nervous system and are therefore necessarily linked within the system in general of subjectivity. (An exception to this is the person who suffers from "blindsight." In this condition the individual will be able to see signifiers, or local elements having simple shapes, but will be unable to transform them into signifieds and so understand what he/she sees.)

On the other hand, the arbitrary bond between signifier and signified results from the conventions of coupling of system types; that is, types of signifiers (English, French, Spanish) generate types of signifieds (concepts produced by those languages). There is, therefore, a double bond between signifiers and signifieds. There is both a necessary and an arbitrary bond, yet because signifiers and signifieds are transformationally distinct, the two are also different. Any isomorphic discussion concerning the differences between signifiers and signifieds is therefore contained within the subjective processes of their integration.

It will be recalled that signifiers are single units of discourse and are prereflectively constructed through a mimetic transformation that for auditory signifiers produces a symmetry of sound with sound image and for visual signifiers a symmetry of percept and visual image. These are necessary closures that differ from the contingent closures of a signified. A contingent closure represents a symmetry of occasion, a coincidence, involving the coupling together of *(a)* a particular order and combination of signifiers (letters of the alphabet) with *(b)* a past or present set of bodily experiences and/or the connection of other signifieds. Simple concepts are therefore the result of a double transformation: first a prereflective, sensorimotor transformation constructing individual units, and then the transformations that come from combining these linguistic units into a combinational form such as words. More complex concepts, evident in sentence constructions, are the result of a series of transformations that on one level combine successive signifiers and on another combine a series of signifieds in an order determined by the conventions of syntax. In other words, the concepts produced by a sentence represent a series of larger recursive transformations that combine the smaller linguistic units (in a process of recursive symmetry) in an order determined by the conventions of syntax. The difference between simple and complex concepts is therefore due to the number and complexity of the recursive transformations that construct successive and overlapping concepts. (The image of concepts presented here is a fractal-type image, where various scales or levels are recursive and self-similarly linked together in a series of spirals of complexity.)

The transformational difference between signifiers and signifieds becomes apparent when we engage with "ambiguous" figures. Psychologists use the phenomenon of ambiguous figures when testing the various features of sensation and perception, and most people are familiar with the visual tricks of such a figure. An ambiguous figure changes as we look at it. One moment it appears as a duck, the next as a rabbit; the profile of a young girl will change suddenly to an old woman; or again, a vase changes into the profiles of two people. But what is transformed? Certainly not the signifiers, the lines, angles, and corners that produce the series of contrasts and distinctions of the drawing. The transformations occur at the point where concepts are formed. Ambiguous figures construct a conceptual "openness" through the ambiguous patterns of the signifiers. This mostly dual pattern provides the occasion in which the normally stable closure of a concept oscillates between two possible closures, that is, between two signifieds.

Ambiguous figures also tell us something about the unconsciousness of closure. Though the patterns of signifiers are ambiguous, a closure we construct at any one time usually excludes the closure produced at another. The alternative concepts are mutually exclusive of the other—we cannot see the duck and the rabbit at the same time. This is the case because in all closures there is a closing off of conscious awareness, a closing off of other possibilities. Put differently, it means that the truth of the rabbit tends to exclude the truth of the duck.

This exclusive effect is important in terms of the nature of rhetoric. For instance, the rhetoric of linguistic signifieds can be such that the conceptual closures produced can seem to defy other possible interpretations. And so life can appear set in an "is-ness" of existence. (The question of rhetoric is discussed in the next chapter.) The benefit of ambiguous figures is that they can point to the differences between signifiers and signifieds. With an ambiguous figure, the signifiers remain the same, while signifieds change before our very eyes.

A further difference between signifiers and signifieds is that concepts, unlike signifiers, are not directly expressible. By this I mean that concepts are not perceptual and therefore play no direct part in a communication transmission between hearer and speaker or between writer and reader. Theirs is not a construction of sensorimotor activity: it is a second-order transformation, and so concepts do not take part in the direct transmission during discourse. Another way of referring to this inexpressibility is to ask how we could express a signified/concept without using signifiers. Direct conceptual expression cannot be demonstrated because it is impossible. In other words, it is not possible to construct words without using letters of an alphabet, just as it is not possible to construct a painting without using a series of contrasting differences of light and shade. Therefore, because concepts/signifieds are second-order emergent constructions, they are not directly involved in the transmission of communication. This distinction simply means that concepts are likely to be more idiosyncratic than signifiers. Yet a given community's sharing a set of concepts (the rule of law, for instance) is not evidence of their expressibility but of conceptual symmetry set against the interpretative differences within that community.

This isomorphic view differs from Saussure's structuralist view of signifieds, which treats them as social elements just as much as signifiers. Both are seen to be linear, and both carry the negative value of difference. For the structuralist, one important difference between signifiers and signifieds relates to the kind of distinctions we make about different languages. For instance, Saussure reports that on one side of the French border the concept of an ox has its signifier *Boeuf* and on the other side *Ochs*.[32] In contrast, an isomorphic view of this example would maintain that signifiers can be similar in different languages (for example, different languages that use the same alphabet) but that signifieds (words), under the arbitrary linguistic habits of different communities, can be different. This dialectical interpretation contrasts with a simpler structuralist view of signifieds. For Saussure, signifieds appear as determining but mysterious agents, as the meaning that is attached to words and therefore as that which determines how language is delimited. Sound images "will correspond to the division along the chain of concepts."[33] This structuralist view completely reverses and changes the emergent order of the isomorphic sign. Such a perspective is only possible if the category of subjectivity is erased from linguistic and interpretative models.

In summary, the distinctions between the second and third levels of the isomorphic sign are produced by virtue of *(a)* their emergent differences, *(b)* their different forms of closure, and *(c)* the inexpressibility of signifieds. The emergent bond between signifiers and signifieds is both arbitrary and necessary, explicit and implicit, and transformational.

Levels of Generality of Signifieds

What are the semantic characteristics of signifieds? In a semiotic sense the term *signified* represents a broad category for a whole range of emergent conceptual transformations. These transformations range in complexity from the more "concrete" linguistic unit such as a "word" to the more abstract unit such as a sentence to highly abstract images and concepts of complexity. Conceptual differences therefore relate to a scale of complexity by which signifieds are recursively combined. The differences of signifieds can therefore be spoken of in terms of levels of abstractness or levels of generality. The more abstract a concept, the greater the number of simple signifieds that have been combined, while the more concrete a concept, the fewer and simpler the combinational character of the signified.

For most linguists the term *signified* usually refers to "the meaning that is attached to words" (Saussure) or just as ambiguously, as Anika Lemaire says, signifies "something in the mind."[34] Lemaire's vague comments do little to advance discussion, and Saussure's notion of meaning is specifically rejected here.

Saussure's rationalist approach identifies meaning with the signified that then determines the nature of signifiers.[35] This identity of meaning and concept is inadequate to account for meaning or for signifieds. We find evidence of this inadequacy in an argument advanced by Jakobson. Jakobson asserted that certain signifiers can exist without a signified. To sustain his argument Jakobson quotes from the novel *Hunger* by Knut Hamsun. The hero has invented the word *kuboa*. "'I have a right, he says, to endow it with whatever meaning I judge appropriate: I do not know yet myself what this word means.'"[36] Jakobson proposes that the signifier *kuboa* has no known meaning; its signified represents a "semantic element of unknown content."[37] From this assertion Jakobson is able to argue that it is possible to have signifiers (words) whose meaning (signifieds) is only the composition of the letters.

From an isomorphic view Jakobson has confused signifier and signified and has relied on Saussure's notion of the signified as meaning. Words are not signifiers, they are signifieds, and the so-called invented word *kuboa* is not a word but a collection of signifiers that cannot be transformed into anything beyond a succession of five letters. The transformation from a collection of letters to a word involves a cardinal symmetry relationship. This relationship of symmetry is one of convention and convenience, but a symmetry necessary for the construction of a word. This symmetry involves a collection of letters with a remembered or present prereflective sensation and/or perception.

If *kuboa* is to become a word, we would have to go through a social and protracted learning process of identifying this collection of letters with a prereflective experience of some kind or with a combination of images generated from other signifieds. For example, we could produce such an identification through a process of learning that coupled the letters *k-u-b-o-a* with anything. We could make up some examples; for instance *(a)* the feeling we get when we eat hot potatoes too fast or *(b)* a view of the Flinders Rangers at twilight or *(c)* the effect of combining a mathematical fifth dimension with the four of perception, and so on. The relationship between the sensory or conceptual experience and the collection of letters is arbitrary, but their identification (which comes from their repetitious synchrony) is necessary (and therefore also motivated). It is necessary in order for a collection of letters like *kuboa* to transform from a collection of letters into being constructed as a word.

Returning to Jakobson's argument, he goes on to question the Saussurian implication that any phonic device such as a phoneme should have the structure of a sign, with both an element of a signifier and one of a signified. Yet what is the signified of a phoneme? As Jakobson asks, "what is the counterpart of a phoneme?" His answer: "Signifier : phonic property; signified ?"[38] From a Saussurian view we can ask what idea or concept can be exchanged for the phoneme. We can also ask the same question of the grapheme *t*. Jakobson's reply to all these questions is that as the phoneme "is not endowed with any specific meaning,"[39] we can have a signifier without a signified.

Yet if the signified is to represent meaning, how is it possible for a phonetic system like speech to have meaning when the cardinal unit on which the system is built is said to have no meaning? If this were the case, all speech would have no meaning. The problem here is the identification of meaning with signifieds. Isomorphically these two are very different. Meaning represents the organizational relationships of cosmic consciousness, subjectivity, and discourse, while signifieds represent a particular set of closures at the third level of the isomorphic sign. Given that meaning is much more general than its derivative, the signified, we can then rephrase Jakobson's conclusions so that phonemes and graphemes do have meaning, but not signifieds.

It has already been argued in this book that phonemes and graphemes have an asynordinate semantic structure consisting of the relations of symmetry, nonsymmetry, and asymmetry. And they emerge prereflectively through symmetry transformations that provide a mimetic projection involving auditory and visual vibration. The meaning of a grapheme and a phoneme is therefore both complex and condensed, unconscious and conscious, implicit and explicit: having an asynordinate structure. In addition, individual phonemes or graphemes do not have a corresponding one-on-one linear relationship to concept/signified. If they did, the relationship between signifier and signified would be a simple extended and linear relationship, and thus the differential complexities of transformation would not be evident. As signifiers do emerge only through symmetry transformations, individual signifiers (both auditory or visual) can exist individually or in groups, such as *kuboa,* without signifieds. Jakobson is therefore correct when he writes a question mark for the signified of a phoneme but incorrect when he says that the phoneme "is not endowed with any meaning."

Concept Formation

A common psychological approach to concepts is to view them as concept formation in children. Lev Vygotsky, like Piaget, described concept formation in children as a function of maturation. In general this developmental approach describes the gradual change that takes place in concepts as a child grows older. This theory proposes that beginning with concrete concepts associated with sensations and perceptions of early childhood, the maturing child gradually develops mastery over more abstract concepts associated with the increasing use of linguistic systems. Vygotsky's developmental approach means there is no binary opposition between the terms "concrete" and "abstract"; rather, these terms represent a sliding scale of generality. He referred to this scale by the phrase "degrees of generality."

> The study of the child's concepts at each age level shows that the degree of generality (plant, flower, rose) is the basic psychological variable according to which they can be meaningfully ordered. If every concept is a generalization, then the relationship between concepts is a relationship of generality.[40]

This fertile approach to concept formation allows for the mixing of elements of concreteness and abstractness within varying degrees of generality. Thus even the most concrete of concepts, like "mum" and "dad," have a degree of abstractness, for each term is a generalization in common use within a linguistic community. Similarly even the most abstract concepts, such as "God," have a concreteness, in that this word is an expression constructed by a particular person on a particular occasion and made up of particular signifiers that are used within a particular syntax.

Vygotsky described his notion of "degrees of generality" as an aspect of the maturation process that is related to the child's ability to generalize from one system to another, or from one set of concepts to another. The mark of an early childhood concept is that it lacks the necessary level of generality to move from the immediate perceptual system to more abstract systems. For instance, as Vygotsky says, "when a child in Piaget's experiments says of one object that it dissolves in water because it was small, and of another that it dissolves because it was big, he merely makes empirical statements of fact that follow the logic of perception."[41]

As the child develops, his/her concepts become more general. An indication of this is when a concept can be transferred out of its original context into another context or system: when, for instance, concepts from structure A (the system of childhood perceptions and the naming of those perceptions) can be transferred to structure B (a system of concepts that incorporates these naming processes and orders them into more general principles). This recontextualization requires a shifting to a plane of greater generality, to a set of concepts that subsumes and governs both A and B.

One of the benefits of this integrating view is its inherent denial that concepts are linear or that they lie in a child's mind "like peas in a bag, without any bond between them."[42] Apart from this stress on integration, however, Vygotsky's study of concept formation in childhood to a large extent employs an unproblematic view of consciousness. Such a view generates some difficulties in its application to adult life. This difficulty comes about because underlying his integrated approach to concept formation is an simple materialism that tends to ignore the often dominant role of rhetoric. Vygotsky's simple approach is described as an ever-increasing system of generalizations that are built upon the generalizations of the preceding levels. Thus the products of intellectual activity of earlier phases are not lost but become the basis of the gradual development of more general and more ordered systems.

This ordering of higher and higher generalizations has some similarity to the complex concepts of sentences, which are the product of the combination of simpler linguistic units. It would be a mistake, however, to equate maturation with sentence construction. The processes of maturation are infinitely more complex than the processes involved in constructing a sentence. Yet it is a simpler ordering process that Vygotsky uses to explain maturation. Thus the abstract generalizations formed in later childhood are seen by him to incorporate, and be structured by, a range of specific, concrete concepts derived from earlier experience. Vygotsky argues that this abstract generalization is lacking in young children because they have no abstract conceptual system that transcends the personal and concrete. The implication of this argument is that maturation should produce linear conceptual systems, which will always incorporate the personal and the concrete.

This equation of maturation with a linear progression is inadequate to account for adult concept formation or for neurotic concept formation in childhood. For example, adults can have conceptual difficulties that arise from a symmetricized consciousness, which is both too abstract and yet too personal. The belief that "I am a failure" is too general yet does not transcend the personal. Such symmetricized concepts would not be possible under Vygotsky's ordered system of maturation. The problem with Vygotsky's approach (to concept formation) is that there are no allowances made for the recursive symmetry of unconsciousness that is inherent to the rhetorical devices of discourse. Nor, as I have implied, does his approach allow for paradigmatic effects on the formation of concepts, that is, the manner in which concepts are modified and constrained by the use of the paradigm of reason.

Vygotsky's maturation model with its "degrees of generality" represents a kind of ideal in concept formation. This is an ideal in which rhetoric is always nullified, in which the generality of terms like *rose* follow on automatically from the more general term *flower,* which in turn follows on to the still more general term *plant.* This particular example of a progression of generality has the benefit of foregrounded semantic connections that stretch from the concrete particular to the more abstract. But Vygotsky's explanation does not account for the relative autonomy that some abstract, rationalist concepts attain through their rhetoric of self-promotion. For example, it does not answer the rhetorical strategies we find in a Platonic view of mathematics, where mathematics is viewed as existing as an ideal form. Nor can it account for the self-evidence of reason or the rhetorical strategies of liberalism, in which the notion of individual autonomy and self-determination represents an ideal form of control. The rhetorical strategies for these relatively autonomous concepts are ones that erases any link to the concreteness of implicit, prereflective meaning and experience.

Vygotsky wrote *Thought and Language* in the 1930s in the political climate of Russian materialism. His theory of concept formation to some extent fits that climate. But in adhering to this simple linear materialism his theory ignored the state of relative autonomy that rationalist concepts, cut loose from their material bases, attain in Western European intellectual traditions. The relatively autonomous nature of concepts is a central and determining theme for Saussure's European linguistics. For Saussure, concepts determine the delimiting nature of language, and therefore the structure of language, as well as the structure of the world. In addition, the structure of the binary sign erases any reference to the concreteness of individual subjectivity. Together, these cardinal erasures ensure a belief in the relative autonomy of concepts as well as the relative autonomy of Saussure's linguistic theory, which was later taken over by poststructuralists such as Derrida. Saussure thus provided an ideal model of language, well fitted to the rationalist idealism of Western European intellectual traditions.

Isomorphically, concepts are not formed in a top-down linear manner, possessing rationalist self-justification, as idealists imply. Nor are they formed as Vygotsky's materialism simply suggests, by ascending levels of generalities. Each of these possibilities, on its own, is inadequate to explain concept formation.

Concept formation is a complex process that has several active ingredients that compete with each other for the dominance of any conceptual system. These active ingredients are, first, the prereflective meanings of the body that involve sensation and perception, and the memory of these concrete experiences that are manifest in signifiers, in simple concepts, and even in some complex concepts. This is a vertical model of ascending levels of generality. For convenience sake we can call it the axis of representation. (This is discussed in more detail in the next chapter.)

In addition to the effects of a vertical set of closures, concepts are also formed by the effects of linguistic transformations that take place within the structure of the sentence, that is, rhetorical effects formed by the syntax. Both these dimensions of discourse, the representational and the rhetorical, are integrated within the unity of every sentence, yet they still compete with each other for conceptual dominance.

Concepts are formed, therefore, not like ascending stairs, as Vygotsky would have us believe, but by the interaction of two different dimensions of discourse, the vertical and the horizontal. And at each level of complexity these closures compete with each other for dominance. We can say that signifiers are dominated by the necessary closure of vibration and image, while simple concepts are dominated by the conventional closure of vocabulary. More complex concepts, however, can be dominated by either the rhetorical closures inherent in the syntax of the sentence, or by the representational closures of felt and learned experience (referred to in the next chapter). In complex concepts either of these forms of closure can reinforce each other or contradict each other. If our rhetoric reinforces the representation of felt experience, the result will be a concept formation that has a high truth value. If, on the other hand, rhetoric is at odds with felt experience, the usual outcome is not ambiguity but a rhetorical dominance. In these cases concept formation will take on an autonomous "reality" and language will be seen as separate from the world.

The dominance of one of these ingredients over the others has been a feature of intellectual life in the West for over two thousand years. Western rationalism has endeavored to erase subjectivity and the implicate order from language, and this endeavor has produced a habit of concept formation by which rhetorical concepts have largely been taken as self-evident truths. Rationalist discourses are therefore dominated by rhetorical truths; witness the rhetorical character of symbolic logic. The reflexivity of different elements of syntax, of rhetoric, is dominant in rationalism because reason awards a primacy to the differences of linguistic units and words over the unconscious symmetry of sensed and felt experience. This primacy operates reflexivity at the level of syntax that thus produces the notion of self-evidence. But self-evidence is actually unnamed, self-reinforcing rhetoric. A feature of the rhetoric of self-evidence is its ability to largely hide its own semantic character as syntactic rhetoric. The success of this Western rhetoric has led to its continuing historical dominance in concept formation.

The historical opposition to the dominance of reason, as a series of ideal concept formations, has come from materialism. But materialism also suffered from its own inability to transcend the bounds of its own reason and, therefore, the rhetorical confinements of its own sentences. Instead of lessening the rhetorical power of syntax, materialism has only provided another kind of rhetorical dominance that is generated from materialist identification patterns. Concept formation in the West therefore has tended toward domination by either materialist or idealist forms of syntactic rhetoric, the effects of which have been to produce simple and mechanistic interpretation of asynordinate systems.

There is no such thing as a materialist or idealist reading of an asynordinate system. Both traditions are unable to adequately account for Bohm's implicate order, the prereflective consciousness, relations as relations of meaning, or the implicit as the implicit, in subjectivity and discourse. These erasures are paradigmatic, and such reasoned perspectives work against acknowledging the dominant and ubiquitous role of rhetoric, particularly the dominating character of their own rhetoric.

In contrast, an isomorphic paradigm does not seek to reestablish the rhetorical concepts of binary oppositions, of idealism and materialism. Rather, an isomorphic paradigm is interested in concept formation as a complex synthesis of different kinds of closures produced by prereflective systems in conjunction with syntactic rhetoric of the sentence. It is the inherent logic of an isomorphic paradigm to give a dominant value to systems of emergence, to the dialects of context and constituent, to the integration of differences, and to the asynordinate and emergent structure that link discourse to subjectivity and cosmic consciousness. What does this mean for concept formation?

This means a very practical and concrete enterprise. It is not so much a new theory, or a denial of rhetoric, but rather the demand that more attention be given to the meaning of the closures of representation. In short, it means providing a prominence to the truths of representation over the truths of rhetoric without denying either. This is the isomorphic approach to concept formation that will be discussed in more detail in the final chapter.

Conclusion

In this chapter I have discussed the two categories of Saussure's sign in terms of the isomorphic sign structure. The similarities and differences between Saussure's sign and the present interpretation are many. While Saussure's sign is different from an isomorphic sign, the significance of a sign as a model for interpreting the world applies to both structures. This means that interpretations can be rationalist, in which case they will adhere to a large degree to Saussure's sign structure, or they can be isomorphic and adhere in some manner to the characteristics of an isomorphic sign.

Chapter 12

Discourse

In this final chapter I want to look at some of the isomorphic features of discourse. Discourse is seen here as the third "layer" of consciousness: a manifest and cultural consciousness that has emerged from the ground of subjectivity, which in turn has emerged from the ground of cosmic consciousness.

The focus of this discussion will not be on traditional categories of linguistic differentiation but on a discussion of the closures of discourse. Recall that a closure represents a completion, a conclusion, or a totalization in which unconscious meaning is closed off from conscious awareness. Closures are a natural and structural feature of all discourse, and our interest in them is not in the possibility of eradicating them (which is impossible) but in describing the strategic role of two major groups of closures. Specifically, I will be concentrating on the closures that determine the two integrated yet distinct axes of discourse: these I have called the vertical axis of representation and the horizontal axis of rhetoric.

The Two Axes of Discourse

What are the two axes of discourse? One way to approach this question is through the difficulties of speech disorders. Roman Jakobson divided all aphasic disturbances into two major types: similarity disorders and contiguity disorders. He made this division on the basis of the dichotomy of the two axes of language. At various times these two axes have been characterized by a variety of terms. Saussure called them the axis of "simultaneity" and axis of "succession."[1] He proposed that the axis of succession had a horizontal, linear dimension and was composed of more than two consecutive linguistic units called syntagms. This axis produced syntagmatic relations in *praesentia*.[2] In contrast, the vertical axis of simultaneities produces associative relations in *absentia*.[3] Isomorphically we can understand the terms *praesentia* and *absentia* as relating to conscious awareness: *praesentia* being present in conscious awareness and *absentia* being unconscious.

Jakobson added to these features with the terms "selection" and "combination," which are the two modes by which language is said to be constructed and understood: he also emphasized the terms "similarity" and "contiguity," the two relations said to underlie the structure of language.[4] In accordance with this view the mode of selection is related with similarity, and combination with contiguity.

The two axes of language therefore have the following features: the horizontal axis combines features; is syntagmatic, successive, linear; and has a contiguity of function. The vertical axis has simultaneity; is selective, associative; and functions on the basis of similarity. Jakobson also added the term "paradigmatic" to the vertical axis.[5] In addition, he divided all aphasic disturbances into the two major types along the lines of the two axes: thus there are similarity and contiguity disorders. Finally, Jacobson proposed that similarity is the basis of metaphor and contiguity the basis of metonymy.[6] We therefore have a very comprehensive description of the two axes. In general, however, we can say that in Jacobson's model the vertical axis has metaphoric dimensions and functions on the basis of similarity, while the horizontal axis functions on the basis of contiguity and can be metonymic.

Now all these different terms fit quite well together and tend to reinforce each other. From an isomorphic perspective these rational descriptions of the two axes of language are interesting, but problematic in that as binary divisions they produce a fractured and divided model, as well as a certain confusion of basic terms.

To take the problem of binary division first: for both Saussure and Jakobson the two axes of language are constructed like dichotomies and relate to each other as if in the shape of a static cross. The dual character of this static cross produces the binary set of terms referred to above, as well as a binary set of linguistic realities. From such a binary perspective Jakobson concludes that there are two types of aphasia: similarity disorders and contiguity disorders. Yet when he describes the empirical evidence of these two kinds of disorder, it does not easily fit the neat binary theory. The main reason for this difficulty is that his basic terms are not sufficiently "pure" in their binary meanings to operate as clearly defined poles opposed to one other. Rather, the meanings of his basic terms *selection* and *combination, similarity* and *contiguity* tend to invade each other to the degree that each is defined in terms of its opposite. In other words, these are integrating and dialectical terms that emphasize different aspects of a synthesis, a synthesis involving the meaning of symmetry and nonsymmetry.

Take, for example, the category of disturbance described by Jakobson as similarity disorders. Impairment of the similarity function would mean a consequent disorder in selection and in the metaphoric dimension of language. *Selection* is a term that indicates a choice between alternatives, which also implies the possibility of substituting one thing for another. Jakobson drew attention to this when he concluded that "selection and substitution are two faces of the same operation."[7] Impairments in the faculty of selection and substitution are thus seen by Jakobson as similarity disorders. However, as implied by Jacobson's comments, the term *selection* carries a more complicated set of meanings than is suggested by the category term *similarity*. *Selection* does not simply imply a singular relationship but rather the similarity of alternatives and therefore the similarity of differences. The rational problem with this intermixing of meaning is that it contaminates any "pure" notion of singularity in aphasic disorders. We find this kind of contamination of meaning also with the terms *combination* and *contiguity*.

According to Jakobson, the combination mode relates to the notion that "any sign is made up of constituent signs and/or occurs only in combination with other signs. . . . [Hence] combination and contexture are two faces of the same operation."[8] Again, here is a description that involves a complex synthesis. The overlap of meaning between the categories of selection and combination is therefore too great to draw the neat binary distinction aimed for by Jakobson. *Selection* and *combination* are themselves overlapping terms, for both deal with substitutions of one kind or another, and both imply the symmetry essential for the integration of differences.

From an isomorphic perspective each axis of discourse functions on the basis of an integration of symmetry, nonsymmetry and asymmetry, which means that selection and combination, similarity and contiguity each has these several features. Each of these four terms tends to indicate a particular emphasis on the relations of either symmetry (similarity, selection) or nonsymmetry (combination, contiguity) but at the same time to imply the reverse characteristics and therefore not to convey the "pure" singularity needed for a binary pair. A binary division is therefore inadequate for describing what appears as two broad dimensions of discourse.

Yet the difficulty of this rational approach does not end there, for the two axes of discourse cannot be seen as a static dichotomy like a cross. If discourse is energetic and dynamic, these two dimensions (which appear more like chaos theory "attractors" than a static cross) will be dynamic, integrated systems that work in harmony with each other. In other words, every metonymy will have some metaphoric aspect and vice versa, or again, every linguistic entity, whether letter, word, or sentence, will be constructed by the integration of the three vertical levels of the isomorphic sign.

In a more general sense this also means that a large amount of linguistic *absentia* (unconscious meaning) will always subtend every *praesentia* (conscious meaning). In other words, discourse will have an asynordinate structure, a structure largely composed of implicate meaning. Specifically this means that as letters, words, and sentences are formed in a horizontal, linear manner, their explicit forms and concepts will have been constructed unconsciously by the vertical dimensions of the isomorphic sign. In this manner discourse can be seen to have two interconnected whorls of meaning: the vertical metaphoric whorl, seen in the structure of the isomorphic sign, and the horizontal metonymic whorl, seen in the linear character of all discourse.[9]

Now these two whorls of meaning, the horizontal and the vertical, can be seen in terms of the symmetrical relationship between constituent and context. For example, Jakobson says that both similarity and contiguity disorders focus on context. Similarity disorders are noted by their reliance on context, while contiguity disorders are always context-deficient. In the *Fundamentals of Language* Jakobson tells us that "in a normal language pattern, the word is at the same time both a constituent part of a superimposed context, the SENTENCE, and itself a context superimposed on even smaller constituents, MORPHEMES . . . and PHONEMES."[10] This description implies that every constituent of discourse always has a context and, in addition, that every constituent can also be a context. Extending Jacobson's description into an isomorphic perspective that views discourse as dynamic, the symmetry transformations that transform constituents and contexts will recursively occur in both whorls of discourse: that is, within the vertical structure of representation (which is also the structure of the sign) and the horizontal dimension of rhetoric (the sentence).

Representation and Rhetoric

Representation

From an isomorphic perspective the two axes of discourse are not characterized by the oppositional terms Jakobson prescribed: similarity and selection versus contiguity and combination. Rather, the two axes refer to the interrelationships inherent in the closures of representation and rhetoric. From this point of view representation functions on the basis of the vertical axis, while rhetoric functions on the basis of the horizontal and linear axis. While representational closures (which I will describe shortly) are different from rhetorical closures, the two axes are nevertheless integrated, so that all discourse will always involve the closures of both representation and of rhetoric.

From an isomorphic perspective the vertical axis of representation can be understood in terms of levels. These levels of subjectivity begin with the prereflective levels of sensations that in turn construct the macrolevels of letter, word, grammatical form, narrative, genre, institution, and culture. In relation to the practice of any particular discourse, this vertical progression of levels will also incorporate a conceptual complexity involving concrete concepts to more general and abstract concepts. Each level is tied to the others by the recursive symmetry of its underlying semantic structure, yet each level will also have different conventional, grammatical, and social features that distinguish it from the other levels.

In chaos theory, scaling represents recursive patterns within recursive patterns, such as water turbulence, which produce smaller whorls within larger whorls within larger whorls and so on.[11] In discourse the same kind of recursive patterns are evident. Yet unlike the systems that chaos theory measures, the recursive patterns of discourse involve self-similar patterns of closure, that is, the repetition of unconscious meaning. Semantically these closures can be spoken of in terms of truth: smaller closures within larger closures, or smaller truths within larger truths that structure the vertical scales of representation. The vertical and representational axis of discourse is therefore not so much a straight-line axis but a vertical spiral of recursive closures.

The effect of viewing discourse in terms of scaling means that the function of the two axes of discourse—representation and rhetoric—will be different at each scaled level of discourse. Yet at each level rhetoric and representation will be an aspect of the other, intertwined together. Depending upon the grammatical construction used, sometimes rhetoric will dominate representation, sometimes this will be reversed. This competition for dominance is not always equally balanced, however. For example, representation will tend to dominate rhetorical effects at the more basic level involving the closure of single linguistic unit or a word. At these levels the function of the horizontal axis will have minimal rhetorical influence because there is a limited linear extension. From an isomorphic perspective, rhetoric represents the intertwining of complex grammatical functions. This being the case, single letters and words have limited grammatical functions.

The "Crisis of Representation"

Some postmodernists believe that "representation" is in a state of crisis. In this respect they seem to have progressed little from early-twentieth-century non-representational painters and artists like the cubists and the futurists, who saw a crisis in representing nature "naturally." From an isomorphic perspective, the postmodern belief in the crisis of representation is understood in terms of the three orientations of consciousness: symmetricization, reason, and isomorphism. Representation can be viewed from each of these three orientations. For example, the naive assumptions inherent in naturalism are consistent with an orientation of symmetricization. From this largely unconscious position nature is seen as unmediated by meaning, observation, discourse or culture, and therefore can be represented innocently, devoid of these mediations because all these meanings are symmetricized into a condensed form of naturalism.

In the social sciences this simple, naturalistic view of "reality" has been successfully challenged, first by early-twentieth-century nonrepresentational painters and later by European language philosophers. This challenge to naturalism was made largely from the orientation of reason. Such an orientation finds distinctions, highlights them, and then overvalues them to an exclusive degree. Such an orientation turned toward truth, representation, art, and literature will find that these are all artificial, consisting of a whole range of distinctions and differences that break the naive web that was supposed to bind us to nature. From such an orientation nature, too, will appear "unnatural," if not fictional.

As I have argued, this orientation of reason has been the basis of the various approaches to discourse by structuralists and poststructuralists. These linguistic philosophers have consistently applied the value of difference (nonsymmetry) to the sign structure and later to identity and to identification. The argument for the crisis of representation has proceeded along the following broad lines: given that difference is the preeminent value, then the process of identifying language must be a process of establishing differences between this phenomenon called language and the unknowable, "real" world. Such a split reduces the closure of representation to a fiction and leaves us with the concepts of surface signifiers (which are the focus of attention) and the unknowable, unnatural "real" world (largely ignored). In contrast with the naive realists of a symmetricized orientation, this rejection of naturalism leads into the antirealist and antihumanist stance of the poststructuralists and postmodernists.

Isomorphically the antirealists are as misled about representation as the naturalists. The dominant relationship in the processes of identifying language, or anything else for that matter, is not nonsymmetry but the integrating effects of symmetry. To identify anything is to produce a closure of at least two complex systems. As closures are constructed by symmetries reinforcing themselves, the dominant relationship at work in all identifications is therefore symmetry. But to value symmetry in this way is not to return to the naturalism of symmetricization. In symmetricization the *only* relationship in representation is symmetry; thus the differential and arbitrary effects of discourse and culture are ignored. From an isomorphic perspective the *dominant* relationship that links the vertical levels of representation is symmetry. This recursive perspective therefore allows for the differential effects of different levels or scales of discourse while maintaining a primary value on repetitive vertical closures. In other words, at every level of discourse the various closures of representation will function slightly differently with respect to the horizontal grammar of that level.

The three perspective paradigms therefore give rise to three different interpretations of representation. A symmetricized perspective will produce the naturalism of the naive realist, a rational view will give us an antirealist stance, while an isomorphic paradigm leads to constructed realism that is a composite of representation and rhetoric.

Representational Closure

At this point I should clarify what constitutes a closure of representation. I have argued earlier that the self-reflection of symmetry (symmetry reinforced) provides the structure of closure and that any closure will produces the unconsciousness of truth. Not all closures are the same, however, for the closures of prereflective consciousness are more closed than the closures of reflective consciousness. The closure of an identity like the letter *t,* constructed prereflectively, is less open to interrogation than the closures of a system like a house or a novel, which have many explicit constituent features. Therefore, in terms of truth not all truths will have the same degree of certainty. They will differ in terms of whether they are constructed prereflectively or are part of a reflective consciousness. A prereflective truth is more certain than a reflective truth because the inhibiting effects of nonsymmetry will be more prominent in reflective consciousness. Therefore the truths of representation will, like all closures, function differently depending upon the level of meaning at which they are constructed and the scale of grammatical complexity with which they interact.

Thus the truths of representation begin at the level of the grapheme and phoneme. These identities are prereflectively and mimetically reconstructed and therefore, as truths, have a high degree of certainty. More complex closures of representation involving sentences will be subject to increasing inhibition and uncertainty from the effects of the many nonsymmetrical relations all sentence exhibits. Offset against the possibilities of this uncertainty will be the syntactic closures of rhetoric that produce propositional certainties. The struggle for dominance between certainty and uncertainty will always make discourse the open site for the uncertainty of many provisional conclusions.

It is clear that a linguistic unit, say the letter *t*, is quite different from the sentence you are reading. Yet they are similar in that they represent forms of representation. Their difference, in terms of representation, is due to their scale of complexity, and it is this difference of scale that produces their different kinds of closure and truth. The letter *t*, for example, has a simple, undisputed mimetic and prereflective truth. Once identified, it is undisputed, for the closure that identifies the letter *t* is a necessary closure, which means that it is not open to interrogation or dispute. The representation inherent in the sentence above, however, is open to many kinds of interrogations—for example, the spelling, syntax, vocabulary, as well as the proposition it contains.

But apart from individual letters of the alphabet, what else in the sentence above constitute the closures of representation? To answer this we need to look at the next level of complexity. The next slightly more complex level of discourse is the construction of words. Words are linguistic contexts where the constituent letters (signifiers) are ordered through integration and distinction in a particular, conventional manner. The context of a word provides a horizontal closure that is not as necessary and prereflective as the identity of a letter. The form of the ordering of the letters is cultural, and the spatial separateness of letters marks differences that are explicitly differential: because of these two factors the closure of words will entail more reflection and less involuntary symmetry than a letter. The closure of words represents a symmetrical construction by convention when the words are spelled correctly. When spelled incorrectly there is no symmetrical construction of convention but a nonsymmetry or nonmimetic representation.

Structuralists and poststructuralists have not been much concerned with the processes involved in the identification of letters and words. These rational theorists were more concerned with the relationship between a word and the signified that it designates.[12] From a rational point of view this relationship is arbitrary and conventional and can be seen by comparing the equivalent French and English words. From an isomorphic perspective this call to convention may be understood as a caution against symmetricization, that is, against naturalizing language. Such a symmetricized view would perceive nouns in particular to have a natural connection with the objects they designate. As rationalists point out, this is clearly not the case. Yet it is also not the case that the *only* connection a noun has to the referent that it designates is a differential and arbitrary relationship.[13] The binary pair of categories, the "natural" and the "arbitrary," are inadequate to account for the integrating relationship inherent in representation. We see this relationship at work in the grammatical function of individual words.

For example, each word in a sentence has a basic grammatical function. Each word functions grammatically as a "part of speech," as a noun, verb, adjective, and so on. This function comes from the dominant symmetrical relationship the word has attained through its use in a linguistic community. Nouns, for instance, produce a vertical and therefore representational closure of word and referent. It is this closure that makes a noun function as a noun. Pronouns, adjectives, adverbs, and verbs construct similar vertical, representational closures of word and referent. If this closure were not constructed, then nouns, pronouns, adjectives, et cetera would have no grammatical function, as is the case with "made-up" words.

Thus all words, to function as words, must refer to something beyond themselves. In this sense, they signify, and they signify mainly through the meanings of symmetry. In this fundamental process of signifying, essential and closed meanings are always produced. These closures are not arbitrary, for the relationship between a word and its referent is not dominated by a reader's reflective consciousness of its arbitrary culture and historical place in the world. This is the very specialized knowledge of etymology possessed by very few. However, even specialists in this field automatically and prereflectively use, for example, nouns as nouns and therefore produce mimetic closures of representation: the symmetries of word and referent.

This necessary closure involving a word and its referent is also found in the case of the first person singular pronoun. When used in its common form, the pronoun *I* has little horizontal dimension but usually a great many representational associations. It is not unusual for *I* to be viewed as symmetrical with the system in general of the authorizing subjectivity. The complexity of meaning in the use of *I* is therefore much more than, say, in the usual use of the letter *t*. The first person singular *I* represents a fusion of letter with name, personal pronoun, and referent in a circular and undifferentiated closure. In this symmetricization, these separate factors are obliterated in the formation of a "nodal" point of condensation and usually repressed meaning.

With so much unconscious meaning fused together within the small explicit symbol of *I*, it was possible for rational philosophers like Jacques Lacan to unravel some of the cultural meanings held captive by its normal symmetricized structure.[14] (It is also interesting to note that Lacan was assisted in carrying out his analysis of *I* by its continual use.) I do not intend to dwell on the differences between the rational interpretations of Lacan and the present perspective. The point I want to make here is concerned with the loading effect, the condensation of representation that this single significant symbol can possess. This condensation of meaning can be viewed as a special symmetricized case of the general vertical nature of representation within discourse.

In contrast to the first person singular pronoun, other pronouns, nouns, adjectives, verbs, and adverbs constitute relatively less complex closures of representation. The scaling factor inherent in these "parts of speech" follows a vertical path that involuntarily and prereflectively identifies first a letter, then a word, and finally a word with referent. Inherent in the operations of a healthy, literate human nervous system is the vertical, recursive linking of meanings within meanings; of the identification of letters and words and grammatical function. Within the confines of a sentence, these closures of representation do not function in a linear mode but, as I have been suggesting, reinforce each other in an interconnected spiral of increasing complexity.

Nonlinguistic forms of representation can also reinforce discursive closures of representation. For example, holding up a pencil to a class of students and saying, "This is a pencil" will produce a symmetry of nonlinguistic representation with a linguistic representation. In other words, the object seen is identified with the pronoun "This" through the contingent and rhetorical closure produced by the verb *to be*.

It follows, then, that the critical question that can be asked of any discourse is the degree to which contingent and rhetorical closures dominate the necessary closures of representation. In other words, how does the symbolic come to dominate felt, sensed, prereflective, and nonsymbolic meaning. Another way of saying the same thing is to ask how we determine how reified or mythological a particular discourse is. To answer these questions, we need to look more closely at the rhetorical aspects of the horizontal whorl of discourse.

Rhetoric

In the isomorphic sense rhetoric operates horizontally through grammatical functions. How do I define rhetoric? From an isomorphic perspective, rhetoric consists of the grammatical construction of contingent closures (truths) that appear (in conscious awareness) as the necessary closures (truths) of representation. Rhetoric therefore always contains a deception by illusion. The deception of rhetoric is to make contingent closures seem necessary, that is, to make imperative the unessential, or to make absolute the relative. The rhetoric of advertising is always strong on making contingent truths appear necessary.

Contingent closures are those that are not necessary for the structural production of a discourse. They can always be opened without damaging the actual processes of expression. Contingent closures function at the level of reflective consciousness, unlike necessary closures, which function prereflectively. They are, therefore, those closures produced by a variety of rhetorical devices on the horizontal axis of discourse. Because rhetoric is constructed contingently, it has the elements of play and deception about it—play, insofar as it pretends contingent truths are necessary; deception, insofar as the make-believe is taken seriously. Those of us who actually believe in the rhetoric we use not only lack a sense of humor but seek to deceive with its use.

The relationship between rhetoric and represented closures will always be dialectical, in that every closure of representation will at the same time function in some manner grammatically and therefore rhetorically. This dialectical relationship means that it is impossible to write or speak a sentence that has no rhetoric and therefore is devoid of deception and illusion. It would be like trying to write a sentence without grammatical construction. This necessary inclusion of rhetoric within the structure of all discourse provides a sense of playful necessity. Playful necessity is perhaps more apparent in the creation of new metaphors in which the synthesis of new meanings is both insistent and unconventional than in the discourses of physics or biology, in which meaning is almost always paraded as exclusively representational.

The sense of a playful necessity may be translated into the conceptual construction of our expressions, but it may not. From a rational perspective, an exclusive either/or alternative operates in this synthesis. If we are rationalists, we may, therefore, choose to ignore the role of rhetoric and instead strive for necessity as the dominant mode of our expressions. Scientists who attempt to map the world mechanically and reductively tend to want to create necessary truths by denying the rhetorical contingency of their utterances. In contrast, some postmodernists choose to celebrate the playful and the rhetorical by disbelieving the necessary truths of representation. Both the scientist who is a naive realist and the postmodernist who is an antirealist focus on only one aspect of the synthesis and erase the other. In a more general social sense, the categories of the serious and the playful are seen by disciplinarians and fundamentalists as exclusive categories; they believe we can choose one or the other, but not both. From an isomorphic perspective, we choose the serious and the playful together, at the same time, to reflect the inherent characteristics of discursive processes. The point of this isomorphic approach is to seek a balance between the truths of representation and the truths of rhetoric. The aim is not to reproduce zero rhetorical effects in our expressions but to expose rhetorical effects as contingent. A rhetorical truth exposed but still used is a mild, playful irony; in other words, it is a serious playfulness.

Myths

To begin to expose rhetorical truths as contingent, we would want to look first at the nature of myth, for myths are usually serious rhetoric (as distinct from serious playfulness). Our first question would therefore be, how does syntactic rhetoric lead to myth? Claude Lévi-Strauss analyzed myth and mythologies from a structural and anthropological perspective. Much of his binary analysis is redundant to an isomorphic view. Some of the characteristics he described, however, are congruent with the isomorphic perspective. For example, he states that "myth is language, functioning on an especially high level where meaning succeeds practically at 'taking off' from the linguistic ground on which it keeps rolling."[15] Agreeing with the tenor of this description, we add that the rhetoric that produces a mythologized discourse creates a set of closures that, while formed by syntax, nevertheless appear to transcend this linguistic ground. Myth reifies meaning by producing its own self-justifying status that tends to erase the direct vertical roots of its own existence. Myth presents itself not as discourse but as "fact," "objective reality," "God given," "laws of nature," "absolute logic" and to some lesser extent as "differance" and "language."

Any discourse can move toward myth by using a rhetorical construction that increases contingent closures at the expense of the necessary closures of representation. A common device to achieve this dominance of rhetorical closure is the fusing together of the closures of representation, not in the manner of a recursive scaling of levels, but as an undifferentiated symmetricized and reified totality. Lévi-Strauss suggest something similar: "If there is a meaning to be found in mythology, it cannot reside in the isolated elements that enter into the composition of myth, but only in the way those elements are combined."[16] From an isomorphic perspective, the combination to which Levi-Strauss refers is a combination by fusing, by symmetricization. In order to construct a myth, therefore, we must combine ordinary elements of representation, not as discrete and constituent elements that form a larger contextual system, but as unconscious members in a reified and undifferentiated totality: a symmetricization of discourse.

The point at which rhetoric "takes off" from representation is very specific to the form of syntax used and is often evident only after close scrutiny of the expression. The general principle involved here, however, is the direction of orientation. Thus, the question to be answered is, "Does the expression produce a top-down or emergent and contextual perspective?" A top-down perspective always comes from a rational and/or a symmetricized consciousness that is manifest in a discourse where rhetorical truths dominate the truths of representation. In other words, top-down perspectives are always created when a priority is given to the second and third levels of the sign. In practical terms this means given to words: signifiers and signifieds. In contrast, an emergent perspective is always produced when the truths of representation dominate the truths of rhetoric. In practice this means when the nonsymbolic has a priority over the symbolic. This is achieved when a discourse points our attention to its own ineffable, nonsymbolic, prereflected, felt, sensed, and intuited meanings. In a contradictory sense, it means when our self-reflective consciousness is directed toward the asynordinate material of its own unconscious and symmetrical foundations.

The Western intellectual tradition has been overwhelmingly a rational, top-down tradition; the truths of rhetoric have conventionally been seen as necessary truths. To understand the mechanisms that produce this unfortunate tradition, it is essential to look at some of the various rhetorical devices that, when combined together, produce the dominance of rhetoric.

Mythic Devices

From an isomorphic perspective, rhetorical closures can be constructed both overtly, in a conscious and positive sense, or covertly, in an unconscious and negative sense.

An overt rhetorical construction will consciously combine at least two meanings as if they were one. This is the field of traditional rhetoric that stretches back to Greek culture in the fifth century B.C.[17] This is not the place to comment at length on this broad field, which has been much written about since Plato and Aristotle. If rhetorical devices function so as to persuade, then the overt rhetorical device will signal to some extent its conscious, persuasive intention, unlike the unconscious rhetorical device. We see this signal of persuasion in the "play" of combining together two explicit meanings in unusual ways by such devices as the metaphor, metonymy, analogy, simile, the joke, rhyme, rhythm, alliteration, and the repetition of sound and intonation. In an overt sense of bringing together two explicit meanings, the metaphor stands out as the exemplary trope. It also has some interesting isomorphic characteristics.

As with Jakobson's model, metaphor has usually been seen as a vertical structure. The double meanings of metaphor appear to fit one over the other in a vertical fashion, as in *the ship of state* or *an open and shut case* or *a sparkling personality.* Despite the appearance of a vertical arrangement of meanings, these metaphors are a linear, horizontal combination of words. Metaphors thus display simultaneously both the horizontal and the vertical whorl of meaning, both a rhetorical, provisional truth and an insistent, representational truth. In addition, this dialectical interaction is achieved at a reflective level of consciousness, where differences are not subsumed by symmetrical closures but maintain their distinctiveness within the overall similarity of the integrated metaphoric system. In "good metaphors" a fine balance is achieved between the integration of differences and their disappearance by choosing overused similarities.

For example, the metaphoric (and dialectical) qualities of *the legs of the table* has disappeared because this trope is so commonplace that its similarities have become too strong for the differences it proclaims. In other words, its similarities have become closed symmetries. For a metaphor to function as a metaphor it should balance differences with similarities (symmetry with nonsymmetry) in a reciprocal playfulness in which neither value is dominant. From a recursive perspective the structure of a metaphor is a good metaphor for discourse.

Overt rhetorical constructions have been of persistent interest to literary critics. More pertinent to the construction of myth are the covert and unconscious grammatical devices and strategies that permeate English syntax. If rhetoric is a means of persuasion, unconscious persuasion is more powerful than conscious devices that we may want to ignore.

Covert rhetoric can come from both symmetricized and rational paradigms. In symmetricized rhetoric syntactic forms can erase qualifications in a fusion that produces the compound identification of myth. This covert rhetoric produces a symmetricization of meaning not unlike the symmetricization of the repressed unconscious. Covert rhetoric represents a form of social repression, a repression of different interests, points of view, and political intent in actions that, in the name of a higher good, suppress the rights of others. Such symmetricized covert rhetoric can come in packages marked *patriotism, God's will, the tribe, the fatherland,* and so on.

But covert rhetoric can also reify by narrowing meaning into exclusive, rational zones that are like islands of data. This rational reification is produced by decontextualization, a process that excludes the integration of contexts. In other words, and as Lévi-Strauss has implied, symmetricized rhetoric tends to fuse together a series of linguistic identities into an unqualified conceptual identification, while a rational rhetoric tends to reify by exclusion. In any practical example, both these forms of covert rhetoric can be found together, reinforcing each other.

In *Language as Ideology* Gunther Kress and Robert Hodge have described a series of syntactic transformations that can be viewed as covert rhetoric.[18] These include such devices as "passive transformation," which has five different categories, "nominalization transformation," which has six different features, and the "negative incorporation," which has two aspects. Each of these syntactic processes functions through deleting or collapsing or simplifying meanings so as to obscure the discreteness of the closures of representation. From a slightly different approach, Kress and Hodge describe these processes of symmetrization and reason as a persuasion intended to continually coerce the reader "into taking the surface form as the real form; and that surface is a radical transformed version of the originally chosen linguistic form."[19]

One of the simplest methods of increasing the power of rhetoric is to combine two of the most commonly used grammatical devices. These two devices are the identificational verb *to be* and the "agent-less phrase." Together they form a mythological certainty that appears beyond interrogation. *To be* manages the rhetorical sleight of hand exceedingly well, particularly in the case where the sentence has a subject–linking verb–noun structure. For instance, the rhetorical necessity inherent in *Jim is sick* or *Jim is a schizophrenic* comes from the way *to be* produces a closure of identity that, within this unqualified form, presents itself beyond interrogation. These closures are rhetorical because while appearing necessary they are only contingent on a particular grammatical form and its use. The identificational verb *to be* constructs a symmetry of subject and object that is absolute, unqualified, and unconscious. The major erasures that have taken place in the above examples are agent and viewpoint and the social and generic conditions within which the statement is made. Once these erasures of meaning are foregrounded, they act as qualifications of the rhetorical truth that had previously been constructed by their erasure.

Similar comments can be made about the agentless phrase. An agentless phrase conceals the author of the discourse in a closure that identifies author with expression. In fiction this form of expression has sometimes been called the omniscient viewpoint. Inherent in the notion of language is the assumption of an omniscient medium that is agentless. In nonfiction the agentless phrase is a common device favored particularly in "objective" television and news reports, in academic writing, and in biblical texts. In these cases the author deletes him/herself from his/her expressions by the expedient of using passive verbs in the third person. The missing identity in these cases is transformed by the semantic identification of expression and author into an "is-ness" of authority and power. Identificational verbs may be combined with the agentless phrase to then increase the seriousness of the rhetoric truth. These are discourses that appear not as dialogue but as transcendent truths without authorship, as if they were handed down from God. The mythological power produced by the combination of an identificational verb and the agentless phrase was such that Michel Foucault once wrote a book about a short sentence with this kind of syntactic rhetoric. It was called *This Is Not a Pipe,* and in it Foucault wrote not about rhetoric but about naturalization.[20]

Decontextualization is another mythic device. Decontextualization refers to a limited focus on a set of significant features that function as if they have no context. Such a selective focus means that the constituent features are reified to an exclusive extent while their related context is erased or ignored. A selected focus can, for example, be constructed when an author omits the social and generic context from his/her writing. This bracketing or decontextualization is common to romantic fiction, but in nonfiction this bracketing becomes an erasure that is sometimes found in the rational traditions of science. For example, the so-called laws of nature are usually decontextualized to the extent that the "laws" are meant to refer only to invariant mechanical and empirical causes and effects. The laws of nature are almost never contextualized so as to refer to that eternal and formative and divine context out of which have emerged the mechanics of cause and effect.

Decontextualization always involves the erasure, in some form, of the context-constituent relationship. This happens when the viewpoint of the author, the genre, or the historical and social setting are disassociated from the rhetorical devices of the expression used. This disassociation can sometimes take the form of a literal reading, in which case words and sentences are not viewed as rhetorical expressions but naturalized into God-given facts beyond the precincts of mere expression. Decontextualization is, however, the natural mythical outcome of a rational perception. For example, in Western academic institutions, this traditional orientation has led to the creation of rational divisions of knowledge that are pursued selectively. The boundaries that divide physics from psychology, or mathematics from linguistics, or literature from science have been predicated on the exclusive separateness of one's own discipline. Because of this selected focus, classical physics and mathematics have had little to say about meaning, discourse, or the processes that mediate and qualify direct observation. Because of this reduced focus, it has been easy to believe that objective reality is separate from and directly accessible to the individual or that mathematics is unmediated by meaning and consciousness.

Similar comments can be made in relation to classical linguistics, mathematics, behavioral psychology, and many other disciplines that focus exclusively on their mainstream concerns. While each of these separate disciplines encompasses different knowledges, they nevertheless share a common orientation and a common set of rationalist decontextual practices and values. These can be summarized in the following manner: (1) belief in the separation of specialized knowledges; (2) a view that situates the individual as separate from the environment and splits mind from body; (3) the practice of establishing "objective reality"; (4) erasure of the qualifications of point of view, time, place, genre, institution, and history from a particular expression; (5) belief that some expressions transcend the material condition of discourse; (6) valuation of distinctions over integration.

The orientation of reason is selective to the extent that it decontextualizes some knowledge based on the value of difference while ignoring knowledge based on recursive symmetry. Such an approach can be very sophisticated, as in Derrida's linguistic philosophy, or very crude, as for example when incidental differences like the color of skin can be overvalued to produce racist discourses. Deconstructive and racist discourses, while very different, similarly decontextualize by overvaluing differences.

In general, mythical discourses have a rhetoric that attempts to persuade us of its inherent seriousness. Its deception is that rhetoric can never be serious, for it is not representation. There are many grammatical mechanisms that can be used to produce the myth of a serious rhetoric, so there are always many symmetricized and rational devices available to the determined persuader.

To use discourse at all is to step into the mire of uncertainty, interpretation, and argumentation. But the tendency to want to produce certainty through discourse is equivalent to the need for ego support through defensive action. The ego is a formation of reason, and to rationalize is to find the appropriate rhetorical devices that are strong enough for self-justification, that is, to reinforce not only the rhetorical truths so they look necessary but to reinforce the ego and make it feel safe from the ignominious torments of uncertainty. As we are unable to express ourselves without using rhetoric, the tendency to push our rhetoric and make it appear necessary is hard for compulsive people to resist. Under these conditions, the problem arises not when we use rhetoric or discourse, but when our rhetorical strategies hide the semantic tendrils of representation and so reverse the natural uncertainty of an emergent and discursive perspective. When this happens, the contingent truths of rhetoric appear as the necessary truths of representation; in the process a myth is created.

Conclusion

In this final chapter we have discussed some of the semantic and symbolic features of subjectivity when manifest as discourse and culture. This discursive area represents the third "layer" of consciousness, which emerges from the potentials of subjectivity, which in turn has emerged from the symmetry potentials of cosmic consciousness or God.

The necessity for the existence of rhetoric in every discourse represents the intellectual "blindsight" of much of Western science and technology. For it is old scientific rhetoric that provides many of the hidden barriers to much new scientific investigation. Take, for example, the instrumental rhetoric that prevents cognitive science from developing into a full science of consciousness. I would suggest that without a training in rhetoric and discourse analysis it is almost impossible to develop an approach to consciousness that is not instrumental and mechanistic. The same can be said of many of the disciplines of science that tacitly rely on the discourse of a deanimated, mechanistic universe. While these discourses may be appropriate for solving mechanistic problems in the explicate order, they fail in the wider field of asynordinate structures, in which animate meaning and consciousness are always operating.

Coda

The theory of consciousness presented here has focused largely on a set of intellectual descriptions of consciousness. This approach provides a comprehensive method and a series of terms for dispassionately tackling some of the many research puzzles related to discourse, culture, society, subjectivity, and cosmic consciousness. In other words, this theory provides a set of workable scientific discourses on consciousness.

This approach is desperately needed in areas like mental health, presently in the grip of the medical model reinforced by a Cartesian view of mind. These mechanistic approaches are unable even to define mental illness beyond the problematics of the norms of tradition supported by a norm-based consensus model developed by committees of the American Psychiatric Association. Neither of these methods is appropriate for a true science of mental health.

Similar broad comments can be made about many social science areas of education, psychology, economics, politics, and linguistics, as well as a range of public-policy areas that deal with the regulation of society. Each of these areas needs to develop concepts and have practices and programs based on a knowledge and understanding of the asynordinate and integrated character of consciousness. For this to happen, the social sciences have to move beyond the present spate of mechanistic and Cartesian responses.

The present intellectual theory of consciousness contrasts with the spiritual or Eastern approach in that it does not describe felt and emotional experience. This is not a belated vote of confidence in introspection, simply the recognition that symmetry relates to feelings and emotions as well as conceptual frameworks.

Now that this book has been completed I seem to have written myself out of the limiting confines of Western intellectualism and will be more confident in the future to also write about aesthetic, emotional, and spiritual aspects of symmetry. One particular result of this escape is the recognition that symmetry produces joy. The joy of symmetry is equal to the type and amount of symmetry experienced. Smaller patterns of symmetry lead to limited joy while greater patterns produce bliss. For example, we can find joy in beautiful mathematics or the symmetry of geometric patterns and mosaics or in works of art. We can experience romantic joy in the identifications of sexual love. But we are overwhelmed with joy in spiritual bliss when we experience the divine.

An experience of the divine is the holistic symmetry of self and God, the implicit connection of the absolute ground of all creation and oneself. Joy and continual happiness are thus to be found along a spiritual path where increasing patterns of symmetry are experienced through reflection, positive attitudes, love, and ego surrender. These are the attributes for which all spiritual aspirants work.

As a basis for consciousness symmetry is beyond the *suchness* and *isness* of the past. The meaning of symmetry is the meaning of Spirit and a panentheistic Spirit is the appropriate basis for this isomorphic theory of cosmic consciousness, subjectivity and discourse.

The present intellectual approach to consciousness, as distinct from the spiritual and Eastern approach, will not bring personal enlightenment, But it does provide a comprehensive method and a set of terms for dispassionately tackling the many research puzzles related to subjectivity and cosmic consciousness. In other words, it begins to provide a workable scientific discourse on consciousness.

This is desperately needed in areas like "mental health," presently in the grip of the medical model reinforced by a Cartesian view of mind. These mechanistic approaches are unable even to define "mental illness" beyond the problematics of the "norms" of tradition supported by a consensus model developed by a committee of the American Psychiatric Association. Neither of these methods is appropriate for a true science of mental health. A science of mental health is quite impossible until there is a science of consciousness, for mental illness can only be defined in terms of the larger and subsuming science of consciousness. I hope the nonrationalist and nonmaterialist theory presented here will be useful for the development of scientific definitions of "mental illness" and "mental health."

Similar general comments can be made about the disciplines of education, psychology, economics, and linguistics, as well as a range of public-policy areas that deal with the regulation of society. Each of these areas needs to develop asynordinate concepts and balanced practices and systems based on a knowledge and understanding of the character of consciousness. For this is to happen, Western science has to move beyond the present spate of mechanistic and

Cartesian responses. Without that development there will be no science of consciousness.

For these reason the isomorphic theory of consciousness developed here needs to be read from the logic of its own premises. It is my hope that through such readings this book will assist in producing productive discussions and further developments in the science of consciousness.

Notes

Introduction

1. Roger Lewin, *Complexity: Life at the Edge of Chaos* (London: J. M. Dent, 1993), 167.
2. See the concluding remarks of "Consciousness Explained, or Explained Away," in Daniel C. Dennett, *Consciousness Explained* (London: Allen Lane, Penguin, 1991), 454–55.
3. John R. Searle, *The Rediscovery of the Mind* (Cambridge, Mass.: MIT Press, 1992), 130.
4. Ibid., 172.
5. Francisco J. Varela, Evan Thompson, and Eleanor Rosch, *The Embodied Mind* (Cambridge, Mass.: MIT Press, 1992), chaps. 3 and 4.
6. Hilary Putnam, *Renewing Philosophy* (Cambridge, Mass.: Harvard University Press, 1992), chap. 1.
7. Francis Crick and Christof Koch, "The Problem of Consciousness," *Scientific American,* September 1992, 111–17.
8. Derek Denton, *The Pinnacle of Life: Consciousness and Self-awareness in Humans and Animals* (Sydney: Allen and Unwin, 1993), 1.
9. David Bohm and B. J. Hiley, *The Undivided Universe* (London: Routledge, 1995), 389.
10. Lewin, *Complexity,* 190.
11. A by-product of this approach is the denial of true randomness. So-called fixed points, as well as any movement of any kind, will always be subject to the structural orderings present within our field of meaning and also present within the environment. This is to say that nothing can escape the ordering principles of symmetry, which are inherent to consciousness and the universe. In this respect, the isomorphic theory agrees with the finding that even chaotic activity always has some form of order.

Chapter 1

1. John Cobb, "God in the Age of Science," an address to the Anglican Church Hunters Hill, Sydney, 3 September 1992.

2. Fritjof Capra, *The Tao of Physics* (London: Flamingo, 1992), 141.
3. Bohm and Hiley, *The Undivided Universe,* 386.
4. Gregory Bateson, *Mind and Nature* (New York: Bantam, 1988), 17.
5. Charles Birch, *On Purpose* (Sydney: New South Wales University Press, 1990), 75.
6. David Bohm and F. David Peat, *Science, Order, and Creativity* (London: Routledge, 1989), 93.
7. David Bohm, *Unfolding Meaning* (London: Ark, 1985), 90.
8. I do not use the term *panpsychism* here or anywhere else in this book when referring to the omniscience or omnipresence of relations of meaning. The reason is that panpsychism is not directly relevant to meaning for it comes out of a rationalist psychology which, amongst other thing, holds to the notions of mentality and units of experience. A more appropriate term in relation to the present emphasis on meaning is *panentheistic.* In this sense, meaning is both immanent and transcendent.
9. *The Concise Yoga Vasistha,* trans. Swami Venkatesananda (Albany: State University of New York Press, 1984), 50.
10. Ibid., 68.

Chapter 2

1. F. David Peat, *Synchronicity: The Bridge between Mind and Matter* (Toronto: Bantam, 1988), 94.
2. Ibid.
3. A. Storr, ed., *Jung* (London: Fontana, 1983), 334.
4. Ibid., 94.
5. Ken Wilber, ed., *Quantum Questions* (Boston: Shambhala, 1985), 51.
6. Paul Davies, "Hawkings Won't Toe the Line," *Australian,* January 17, 1996.
7. Heinz R. Pagels, *Perfect Symmetry* (New York, Bantam, 1991), 170–205.
8. B. Gruber and R. S. Millman, eds., *Symmetries in Science* (New York: Plenum Press, 1979).
9. Pagels, *Perfect Symmetry,* 178.
10. Joe Rosen, *Symmetry Discovered* (London: Cambridge University Press, 1975), 5–6.
11. Ibid., 96.
12. Heinz R. Pagels, *The Cosmic Code* (London: Penguin, 1994), 315.
13. Ibid., 304.
14. Wilber, *Quantum Questions,* 51.
15. Ibid., 80.
16. Ibid., 170.
17. Rupert Sheldrake, *The Presence of the Past* (London: Fontana, 1988), 97.
18. Ibid., 108.
19. Ibid.
20. *The Upanishads,* trans. Swami Prabhavananda and Frederick Manchester (Hollywood, Calif.: Vedanta, 1983), 73.
21. Storr, *Jung,* 343.
22. Ibid., 341.
23. Ibid., 342.
24. Ibid., 339.

25. Stanislav Grof, *The Holotropic Mind* (San Francisco: Harper, 1990), 178.
26. Wilber, *Quantum Questions,* 202.
27. Ibid., 10.

Chapter 3

1. These etymological references are taken from the *Oxford English Dictionary,* 2d ed.
2. N. Katherine Hayles, *Chaos Bound: Orderly Disorder in Contemporary Literature and Science* (London: Cornell University Press, 1990), xiv.
3. Wilber, *Quantum Questions,* 27.
4. Pagels, *The Cosmic Code,* 284.
5. Henry Margenau, *The Nature of Physical Reality* (Connecute: Woodbridge: Conn., Ox Bow Press, 1977), 428.
6. Peat, *Synchronicity,* 16.
7. *The Concise Yoga Vasistha,* 50.
8. David Bohm, *Wholeness and the Implicate Order* (London: Ark, 1983), 149.
9. Basil J. Hiley and F. David Peat, eds., *Quantum Implications: Essays in Honour of David Bohm* (London: Routledge, 1991), 43.
10. David Bohm and Basil J. Hiley, *The Undivided Universe* (London: Routledge, 1995), 389.
11. John Hogan, "Last Words of a Quantum Heretic," *New Scientist,* 27 February 1993, 38–42.
12. Capra, *The Tao of Physics,* 353.
13. *The Concise Yoga Vasistha,* 51.
14. Ibid., 150.
15. Ibid., 149.
16. Bohm, *Wholeness,* 178.
17. Ibid., 199.
18. Sigmund Freud, *The Interpretation of Dreams,* ed. and trans. James Strachey (Middlesex: Penguin, 1986), 773.
19. Bohm, *Wholeness,* 210.
20. Bohm and Peat, *Science, Order, and Creativity,* 104.
21. Bohm, *Wholeness,* 208.
22. Gordon D. Globus, "Three Holonomic Approaches to the Brain," in Hiley and Peat, *Quantum Implications,* 373–85.
23. Bohm and Peat, *Science, Order, and Creativity,* 185.
24. The abstract description here of the relations of symmetry, nonsymmetry, and asymmetry is based largely on chapter 13 of Alfred Korzybski's *Science and Sanity* (Lakeville, Conn.: Institute of General Semantics, 1948), 188–203.

Chapter 4

1. Humberto Maturana tells us that the nervous system is a closed, "autopoietic" system. Autopoiesis refer to the concepts of self-production, self-perpetuation, and self-containment: in simple terms, the "closure" of living systems. A closed system implies boundaries to an internal organization where transformations can occur. A

closed system therefore refers to an organizational closure. Organizational closure in turn implies autonomy and the generation of autonomous behavior. In 1972 Maturana began to use the term "autopoiesis" to refer to the "circular organisation" that characterizes living systems. See Humberto Maturana and Francisco J. Varela, *Autopoiesis and Cognition: The Realisation of the Living* (Dordrecht, Holland: Reidel, 1980), xvii. Since then the term autopoiesis has been used increasingly to refer to the recursive dynamics that produce the autonomy of living systems. The concept of autopoiesis has been largely constrained by the context of its biological formulation, even though autopoietic properties have been seen to operate in social formations, while some writers interested in computers and artificial intelligence have used the concept in that field. For comments in both these areas, see Milan Zeleny, "Autopoiesis: A Paradigm Lost?" in *Autopoiesis, Dissipative Structures, and Spontaneous Social Orders*, ed. Milan Zeleny, AAAS Selected Symposia Series (Boulder, Colo.: Westview Press, 1980), 3. Also see Terry Winograd and Fernando Flores, *Understanding Computers and Cognition* (Reading, Mass.: Addison-Wesley, 1988). Maturana and Varela view the environment as a "trigger" and a source of "perturbations." See Francisco J. Varela, *Principles of Biological Anatomy*, vol. 2 (New York: North Holland, 1979), 262. Both these terms imply that the environment has a marginal importance to the transformations that take place within any system.

2. Varela, *Principles of Biological Anatomy*, 245.
3. Ibid., 276.
4. Zeleny, "Autopoiesis," 5.
5. Maturana, "Autopoiesis: Reproduction, Heredity, and Evolution," in Autopoiesis, Dissipative Structures, and Spontaneous Social Orders, ed. Milan Zeleny, AAA Selected Symposia Series (Boulder, Colo: Westview Press, 1980), 52–53.
6. Varela, *Principles of Biological Anatomy*, 261.
7. Ibid., chap. 15, and Maturana and Varela, *Autopoiesis and Cognition*.
8. Maturana, *Autopoiesis and Cognition*, 57.
9. Ibid., 120 (emphasis added).
10. Varela, *Principles of Biological Anatomy*, 243.
11. Ibid., 238.
12. As in ibid., 250.
13. Ibid., 250.
14. Ibid., 254.
15. Ibid.
16. Ibid., 255.
17. Ibid., 262.
18. From a rational perspective, such a definition of a necessary truth must also represent the definition of a necessary illusion, for an illusion is produced when two distinct things are identified together as if they were one. From an isomorphic perspective, the symmetry of a prereflective consciousness is central to any analysis and therefore is consonant with the statement that objects are symmetrical with the percepts we construct of them. If we want to call this symmetry an illusion, then illusion is necessary for perception to function.
19. Monika M. Langer, *Merleau-Ponty's Phenomenology of Perception* (London: Macmillan, 1989), 97.

Chapter 5

1. Interview by Renee Weber, in Hiley and Peat, *Quantum Implications,* 436.
2. Ibid., 443.
3. Bohm, *Unfolding Meaning,* 87.
4. A summary of his position is found in Umberto Eco, M. Santambrogio, and P. Violi, eds. *Meaning and Mental Representations* (Bloomington: Indiana University Press, 1988).
5. Emile Benveniste, *Problems in General Linguistics,* trans. M. E. Meek (Coral Gables, Fla.: University of Miami Press, 1971), 106.
6. For a critique of I. A. Richards's use of meaning and the ambiguities of his view, see T. C. Pollock, *A Theory of Meaning Analysed: A Critique of I. A. Richards' Theory of Language and Literature,* General Semantics Monographs, no. 111 (Lakeville, Conn.: Institute of General Semantics, 1961).
7. A. J. Greimas and J. Courtes, *Semiotics and Language: An Analytical Dictionary,* trans. Larry Crist et al. (Bloomington: Indiana University Press, 1982), 187.
8. Charles Morris, *Writings on the General Theory of Signs* (The Hague: Mouton, 1971), 95.
9. Greimas and Courtes, *Semiotics and Language,* 187.
10. Ludwig Wittgenstein, *Philosophical Investigations,* trans. G. E. M. Anscombe (Oxford: Basil Blackwell, 1989), 20.
11. Ibid., 14.
12. Gunther Kress, ed., *Halliday: System and Function in Language, Selected Papers* (Oxford: Oxford University Press, 1981), xi.
13. Colin McGinn, *Wittgenstein on Meaning* (Oxford: Basil Blackwell, 1987), 1.
14. T. E. Hill, *The Concept of Meaning* (London: George Allen and Unwin, 1974), 1.
15. For a comprehensive review of the different ways in which many writers have viewed meaning, see K. L. Pike's chapter on meaning in *Language in Relation to a Unified Theory of the Structure of Human Behavior* (The Hague: Mouton, 1967), 598–640.
16. Ferdinand de Saussure, *Course in General Linguistics,* ed. Charles Bally and Albert Sechehaye, trans. Wade Baskin (Glasgow: Fontana/Collins, 1978), 120.
17. Jacques Derrida, *Margins of Philosophy,* trans. Alan Bass (Chicago: University of Chicago Press, 1982), 3–27.
18. Jacques Derrida, *Speech and Phenomena,* trans. David B. Allison (Evanston, Ill.: Northwestern University Press, 1989), 140–41.
19. Ibid., 133.
20. Ibid., 134.
21. Jacques Derrida, *Of Grammatology,* trans. Gayatri Chakravorty Spivak (Baltimore: John Hopkins University Press, 1980), 7.
22. Keith Allan, *Linguistic Meaning,* vol. 1 (London: Routledge, 1986), chap. 2.
23. Noam Chomsky, *Language and Mind* (New York: Harcourt Brace Jovanovich, 1972), 104.
24. Ibid., 17.
25. Ibid., 18.
26. Roger Fowler, *Linguistics and the Novel* (London: Methane, 1979), 12.

27. Martin Heidegger, *Being and Time,* trans. J. Macquarrie and E. Robinson (Southampton: Basil Blackwell, 1983), 193.
28. See, for instance, Karl Popper's comments in Popper and Eccles, *Self and Its Brain,* 94; and Searle, *Rediscovery of the Mind,* 172.
29. Freud, *The Interpretation of Dreams,* 773.
30. J. Laplanche and J.-B. Pontalis, *The Language of Psycho-analysis,* trans D. Nicholson-Smith, (New York: Norton, 1973), 235.
31. Ibid., 243.
32. Freud, *The Interpretations of Dreams,* 381.
33. Ludwig Wittgenstein, *Tractatus Logico-Philosophicus,* trans. D. F. Pears and B. F. McGuinness (London: Routledge and Kegan Paul, 1961), 151 (T. 4. 1212).
34. Benveniste, *Problems in General Linguistics,* 107.
35. Ibid., 302.
36. Saussure, *Course in General Linguistics,* 120–21.
37. For a critical analysis of Saussure's view of language not in use, see A. Freadman, "On Being Here and Still Doing It," in *The Foreign Bodies Papers* (Sydney: Local Consumption, 1988),
38. I do not use *seme* in the sense of a minimal unit of meaning. Nor is this term used because of its history as an analytical and differential form. The general notion of a unit, if used on its own, produces an "objective" view of the material being investigated. In contrast, the present isomorphic view stresses the transformative character of units.
39. Greimas and Courtes, *Semiotics and Language,* 191.
40. Saussure, *Course in General Linguistics,* 80.
41. Roman Jakobson and Morris Halle, *Fundamentals of Language* (The Hague: Mouton, 1980), 95.
42. Ibid., 83.
43. Ibid., 84.
44. Greimas and Courtes, *Semiotics and Language,* 191.
45. "Metaphor, therefore, is determined by philosophy as a provisional loss of meaning." Derrida, *Margins of Philosophy,* 270.
46. Nietzsche writes about recognizing untruth as a condition of life that places his philosophy beyond good or evil. *Beyond Good and Evil,* trans., R. J. Hollingdale (Middlesex: Penguin, 1987), 17–18. The present argument for the apparent solidity of the linguistic unit also goes beyond the binary values of truth or falsity. The apparent solidity is not considered as an error, however, but as a necessary condition of consciousness. It is thus not a question of untruth but of the conditional truth of appearances.
47. Benveniste, *Problems in General Linguistics,* 108.
48. John R. Searle, *Speech Acts* (Cambridge: Cambridge University Press, 1985), 16.

Chapter 6

1. Chomsky, *Language and Mind,* 27.
2. See, for instance, Karl Popper's comments about the uselessness of discussing questions such as "What is consciousness" in Popper and Eccles, *Self and Its Brain,* 100.

Another view of consciousness is contained in a comparison between minds and computers by Penrose, *The Emperor's New Mind.* Penrose chooses to define consciousness ambiguously as "the 'seeing' of a necessary truth" (577). Yet another position on consciousness common in philosophy is the one put foreword by Martin, *Shadows in the Cave.* Martin tells us that we should not be worried "about our inability to define consciousness in words," for "pure consciousness, as the mystics point out, is itself featureless" (65–67).

3. Laplanche and Pontalis, *The Language of Psycho-Analysis,* 84.
4. Ignacio Matte-Blanco, *The Unconscious as Infinite Sets* (London: Duckworth, 1975), 96.
5. Ibid., 60.
6. Ibid., 96.
7. Ibid., 109.
8. Ibid., 97.
9. Ignacio Matte-Blanco, *Thinking, Feeling, and Being,* ed. David Tuckett (London: Routledge, 1988), introduction, 23.
10. Matte-Blanco, *Unconscious as Infinite Sets,* 96.
11. Ibid., 150.
12. Ibid., 42.
13. Laplanche and Pontalis, *The Language of Psychoanalysis,* 82–83.
14. Sigmund Freud, "The Unconscious," in *Metapsychology: The Theory of Psychoanalysis,* trans. James Strachey (Middlesex: Penguin, 1984), 190.
15. Ibid., 191.
16. Matte-Blanco, *Thinking, Feeling, and Being,* 16.
17. Ibid.
18. Matte-Blanco, *Unconscious as Infinite Sets,* 96.
19. Ibid.
20. Ibid., 97.
21. Ibid., 14.
22. Ibid., 151. In this reference Matte-Blanco argues for a difference between symmetry and asymmetry and the psychic states of consciousness and unconsciousness. He says, "The distinction between conscious and unconscious does not coincide with the distinction between symmetrical and asymmetrical." Yet Matte-Blanco's major proposal is that there is a coincidence between the psychic and these relations. This denial seems due to the conventional problem of the relationship between the conscious and unconscious. By accepting the traditional Freudian view that the unconscious can be known only indirectly, Matte-Blanco arrives at the contradictory situation of accepting a coincidence of psychic states and these relations, yet explicitly denying this coincidence when addressing the question of the relation between conscious and unconscious. He suggests rather that this question "can, perhaps, be examined better in terms of symmetrical and asymmetrical relations." Thus he arrives at the "possibility of gradual transitions, of a continuum from asymmetrical to symmetrical." He asserts that things must be either conscious or unconscious, not both, but in relationships there can be a gradual transition. I suggest Matte-Blanco's unsatisfactory conclusion comes about through his rationalist perspective, which separates psychic states one from another, but is directly undermined by his isomor-

phic focus on relationships, which will always produce not separations but connections.

23. A linguistic objection to this argument could rest on the distinction between the process of representing and the item represented. This line of argument would run as follows: while the process of representing is a conscious one, the unconscious processes represented (symmetrical relations) are not. In other words, ordinary language is conscious, whereas dreams, slips of the tongue, jokes, et cetera, are not ordinary language but a kind of subterranean eruption of unconscious forces into language. Such is the conventional Lacanian view. The problem with this linguistic objection is that it contains the traditional commitment to the separation of the conscious and the unconscious, since the distinction between "representing" and the "represented" coincides with that between conscious and unconscious, which means that we cannot know the unconscious consciously. This view of language is rationalist and formalist, language as "words on the page" from which we, and our perceptual processes, are separate. In this set of discriminations, there are three separate entities: language, consciousness, and implied unconsciousness.
24. Relevant to the question of the "balance" of consciousness to unconsciousness in any statement is Freud's view that the "unconscious is the larger sphere, which includes within it the smaller sphere of the conscious." *The Interpretation of Dreams,* 773.
25. Ibid.
26. Freud, "The Unconscious," 192.
27. Matte-Blanco, *Unconscious as Infinite Sets,* 104.
28. Freud, *The Interpretation of Dreams,* 773.
29. Ibid., 87.
30. Ibid., 91.
31. Laplanche and Pontalis, *The Language of Psychoanalysis,* 85.
32. Ibid.
33. Matte-Blanco, *Unconscious as Infinite Sets,* 111.
34. Matte-Blanco, *Thinking, Feeling, and Being,* 23.
35. Umberto Eco, *The Role of the Reader* (London: Hutchinson, 1981), introduction, 3–43.
36. This may seem a strange use of the word *known,* but we would rather use *known* in relation to unconscious meaning than such terms as *felt* or *intuited.* This is because, paradoxically, it is the unconscious that deals in certainties (as Freud tells us, there are no degrees of certainty in the unconscious). The word *known* therefore implies certainty more than does *felt* or *intuited.*

Chapter 7

1. Thomas S. Kuhn, *The Structure of Scientific Revolutions,* 2d ed. (Chicago: University of Chicago Press, 1970), 6.
2. Ibid., 82.
3. Hayles, *Chaos Bound,* 1.
4. I have borrowed these three terms, symmetricized, rational, and isomorphism, from

different places. The concept of a symmetricized consciousness comes from the psychoanalytic work of Ignacio Matte-Blanco, who applies it generally to the organization of unconscious phenomena. The term *rational* has its traditional sense. However, in using it to define a paradigm of consciousness I will be restricting it more than do writers like David Bohm. Along with most others, Bohm sees reason as a privileged and universal thinking process. But Bohm extends the usual meaning of reason so as to include intuition. See Bohm and Peat, *Science, Order, and Creativity* 145–50. I see intuition as an aspect of an isomorphic consciousness, and not an element in a reasoned consciousness. In addition, rationalism and empiricism are traditionally distinct, in part because of their histories, but also because of their practices. However, from the point of view of consciousness, empiricism falls within the paradigm of reason because its central semantic value is difference, that is, the differences inherent in sense perceptions. The term *isomorphism,* as I have already said, comes from Katherine Hayles.

5. Langer, *Merleau-Ponty's Phenomenology,* part III, "The Cogito."
6. See the introduction to Matte-Blanco, *Thinking, Feeling, and Being.* Matte-Blanco uses the term "symmetrization" to indicate when items in a collection are regarded subjectively as the same in all respects. In this sense symmetrization represents the function of the Freudian unconscious.
7. Wilhelm Reich described in some detail what amounts to the symmetricization of discourse and thinking in *The Mass Psychology of Fascism,* trans. V. R. Carfagno (New York: Farrar, Straus and Giroux, 1970), in particular chap. 11.
8. Jacques Derrida, *Writing and Difference,* trans. Alan Bass (Chicago: University of Chicago Press, 1978), chap. 10.
9. For instance, the structuralist difficulty over the way in which difference relates to the linguistic sign.
10. On the connection of schizophrenia and rationalism see Louis A. Sass, *Madness and Modernism* (New York: Basic Books, 1992).
11. See the discussion in chapter 9.
12. Sheldrake, *Presence of the Past,* 109.
13. Ammarchi is a divine soul, presently living as a Hindu spiritual leader in India. Her devotees call her "Amma" or "Mother." Her spiritual name is Mata Amritanandamaya. Each year Amma tours countries in Europe, North America, and the Asian Pacific region, giving blessings to her followers and devotees.
14. James Gleick, *Chaos: Making a New Science* (London: Cardinal, 1989), 61.
15. Ibid.
16. Ibid., 226.
17. Ibid.
18. Ibid., 115.
19. I. Stewart, *Does God Play Dice: The Mathematics of Chaos* (Oxford: Basil Blackwell, 1989), 196.
20. Hayles, *Chaos Bound,* 156.
21. N. Katherine Hayles, "Chaos as Orderly Disorder: Shifting Grounds in Contemporary Literature and Science," *New Literary History* 20 (1990): 308.

Chapter 8

1. John P. Frisby, *Seeing Illusion, Brain, and Mind* (New York: Oxford University Press, 1979), 26.
2. Karl H. Pribram, *Languages of the Brain: Experimental Paradoxes and Principles in Neuropsychology* (New York: Brandon House, 1977), 132.
3. Popper and Eccles, *Self and Its Brain,* 31.
4. Rosen, *Symmetry Discovered,* 4.
5. Chomsky, *Language and Mind,* 17.
6. Gleick, *Chaos,* 86.
7. Mark Johnson, *The Body in the Mind* (Chicago: University of Chicago Press, 1987), 171.
8. Rupert Sheldrake, *The Rebirth of Nature* (London: Rider, 1990), 118.
9. Derrida, *Margins of Philosophy,* 16.

Chapter 9

1. Popper and Eccles, *Self and Its Brain,* vii.
2. U. T. Place, "The Brain and Consciousness," in *Introductory Readings of Metaphysics,* ed. R. Taylor (Englewood Cliffs, N.J.: Prentice-Hall, 1978), 86–94.
3. Place seems to suggest that "the is of definition" represents the kind of identity that Leibniz referred to as "the identity of indiscernibles." This is Albert Shalom's reading of Place's definition, with which I concur. See Albert Shalom, *The Mind/Body Conceptual Framework and the Problem of Personal Identity* (Atlantic Highlands, N.J.: Humanities Press International, 1989), 93–111.
4. Place, "The Brain and Consciousness," 90.
5. Ibid., 94.
6. Ibid. Place attempts to fill in the blanks between these two kinds of descriptions. Shalom reports that Place in the early 1960s added to his original hypothesis by proposing that the identity of mind and brain processes was an identity of mind processes and the *micro*processes of the brain. As Shalom states it, "The point of this identity thesis is, then, that a distinction must be made between 'processes' and 'micro-processes'" (*Mind/Body Conceptual Framework,* 102). Shalom's criticism of this hypothesis is that the suggested identity of mind and body is now too unlike the identity of the other examples used by Place. Shalom's argument runs in the following manner: If consciousness is to be regarded as a "process" to which the brain's "microprocesses" correspond, then "consciousness" is not an observable process in the sense that lightning or a cloud are observable. From Shalom's traditional point of view, we can only infer a person's state of mind, his or her desires, emotions, and intentions; thus consciousness is a nonobservable that Place asks us to identify with the observables of brain activity. Shalom sums up his criticism of Place thus: If I consider someone else, I may be able to observe the person's "brain processes," but I cannot observe his or her own actual experiences of "consciousness." And therefore I cannot establish any simple or straightforward "identity" between the brain processes I can in principle observe, and what the person is actually experiencing (107). While Shalom is content with this criticism of Place's identity theory, it does

not appear require a substantial change in Place's basic hypothesis. Shalom's criticism does, however, bring to our attention the need to develop descriptions of the processes of consciousness that are isomorphic with what we know about the micro-processes of brain activity. The need for an isomorphism of descriptions is reflected in the evidence of it in Place's own examples of "a cloud" as a "mass of tiny particles in suspension" and also with "lightning" as "a motion of electric charges." While both sets of terms in these examples are different from each other, they are not so different as to exclude one another. There is nothing in the description "a flash of lightning" contrary to the second description, "a motion of electric charges." But more than this, both statements reinforce the other isomorphically, as do the two descriptions of clouds.

7. Ibid., 110.
8. Popper in Popper and Eccles, *Self and Its Brain,* 96.
9. Ibid., 97.
10. Popper in Popper and Eccles, *Self and Its Brain,* 82.
11. Popper and Eccles, *Self and Its Brain,* 361.
12. Ibid., 361.
13. Ibid., 362.
14. John McLeish, *Soviet Psychology* (London: Methuen, 1975), 89.
15. See McLeish's comments, ibid.
16. George A. Millar, *Psychology: The Science of Mental Life* (London: Hutchinson, 1966), 175.
17. Popper and Eccles, *Self and Its Brain,* 94.
18. Ibid., 67.
19. Ibid., 191.
20. Harry K. Wells, *Pavlov and Freud: I Ivan P. Pavlov Towards a Scientific Psychology and Psychiatry* (London: Lawrence and Wishart, 1956), 64–100.
21. Ibid., 99.
22. Ibid., 64. In respect to Pavlov's interest in psychology, it is interesting to note that he started off his scientific career strongly disapproving of psychology; only in later years, in his psychiatric research, did he change to the degree that he called himself an "experimental psychologist." It is clear, however, that Pavlov was primarily a physiologist.
23. Maturana and Varela, *Autopoiesis and Cognition,* 119.
24. Ibid., 46.
25. McLeish, *Soviet Psychology,* 69.
26. The term "shunt off" is used by Varela to describe the influences of inhibition in *Principles of Biological Anatomy,* 240.
27. J. Bray et al., eds., *Lecture Notes on Human Physiology (*Oxford: Blackwell Scientific Publications, 1986), 32.
28. See chapter 2 of Patricia S. Churchland, *Neurophilosophy* (Cambridge, Mass.: MIT Press, 1986).
29. J. deGroot and J. G. Chusid, *Correlative Neuroanatomy* (New York: Prentice-Hall, 1985), 22–31.
30. Peter K. Anokhin, *Biology and Neurophysiology of the Conditioned Reflex and Its Role in Adaptive Behavior,* trans. S. A. Corson et al. (Oxford: Pergamon, 1974), 503.

31. Ibid., 509.
32. Ibid., 272.
33. Ibid., 502.
34. Ibid., 293–94.
35. McLeish, *Soviet Psychology,* 79.
36. Anokhin, *Conditioned Reflex,* 294.
37. Langer, *Merleau-Ponty's Phenomenology,* 29.
38. Ibid., preface.
39. Ibid., xv.
40. Ibid., 118. On this point Langer refers us to Merleau-Ponty's own words: "reflection does not itself grasp its full meaning ["sens"] unless it refers to the unreflective third fund of experience which it presupposes, upon which it draws, and which constitutes for it a kind of original past."
41. Bray et al., *Notes on Human Physiology,* 32.
42. Ibid., 33.
43. Pribram, *Languages of the Brain,* 85.
44. Millar, *Psychology,* 173–92.
45. de Groot and Chusid, *Correlative Neurophysiology,* 330.
46. Langer, *Merleau-Ponty's Phenomenology,* 39.
47. Ibid., 40.
48. Ibid., 162.
49. See, for instance, Eccles' explanation in Popper and Eccles, *Self and Its Brain,* 330–33.
50. Ibid., 363.
51. Denton, *The Pinnacle of Life,* 11–16.
52. Ibid., 15.
53. Ivan V. Pavlov, *Lectures on Conditioned Reflexes,* vol. 1, trans. W. Horsley Gantt (London: Lawrence and Wishart, 1963), 221.

Chapter 10

1. Saussure, *Course in General Linguistics,* chap. 3, "The Object of Linguistics."
2. J. Aitchison, *The Articulate Mammal* (London: Hutchinson, 1976), chap. 9.
3. B. Klob and I. Q. Whishaw, *Fundamentals of Human Neuropsychology,* 3d ed. (New York: W. H. Freeman, 1990), 334.
4. Ibid., chap. 21, "Memory." The problem of naming areas of the brain involved in conceptualization appears to be similar to that identifying memory areas. Conceptualization involves an integration of both memory and sensory meaning, and many areas of the brain are involved in these functions. In addition, conceptualization or the production of signifieds varies in complexity, depending upon the nature of the concept. For instance, the complexity of an image that comes from the identification of a simple object like a table or the letter *a* is much less than the complexity of the conceptualization of some system or theory. We could expect, then, that different areas of the brain are involved in different types of conceptualization.
5. Ibid., 155–56.
6. These rational values can relate to a range of activity, such as parental responses to

their children. Alice Miller is particularly interesting in her descriptions of the types of rationalizing upbringing many of the German SS officers were subject to in their early childhood years. Miller does not use the term *rational,* but the child-rearing values she describes are isomorphic with the values of what I would suggest are a rational approach to childhood. See Alice Miller, *For Your Own Good,* trans. H. and H. Hannum (New York: Noonday Press, 1991).

7. Maurice Merleau-Ponty, *Consciousness and the Acquisition of Language,* trans. Hugh Silverman (Evanston, Ill.: Northwestern University Press, 1973), 11.
8. Jacques Lacan, *Ecrits: A Selection,* trans. Alan Sheridan (London: Tavistock, 1985), 65.
9. Jean Piaget, "Language and Thought from a Genetic Point of View," in *Language in Thinking,* ed. P. Adams (Middlesex: Penguin, 1973), 170–79.
10. Julia Kristeva also points to the connection between such presymbolic gestures as laughter and the sentence in *Desire in Language,* ed. Leon Roudiez, trans. Thomas Gora, Alice Jardine, and Leon Roudiez (New York: Columbia University Press, 1980), 271–91.
11. Ibid., 286.
12. Derrida, *Of Grammatology,* 158.
13. Kristeva, *Desire in Language,* 287.
14. Langer, *Merleau-Ponty's Phenomenology,* 111.

Chapter 11

1. See Antonio R. Damasio and Hanna Damasio, "Brain and Language," *Scientific American,* September 1992, 63–71.
2. Stanley Fish, *Doing What Comes Naturally* (London: Duke University Press, 1989), 83.
3. Saussure, *Course in General Linguistics,* 120.
4. Ibid., xix.
5. Ibid., 117.
6. Ibid., 6.
7. Ibid., 8.
8. Ibid., 13.
9. Ibid.
10. Ibid., 38.
11. Ibid., 104.
12. Roman Jakobson, *Six Lectures on Sound and Meaning* (Cambridge, Mass.: MIT Press, 1978), 11.
13. Ulric Neisser, *Cognitive Psychology* (New York: Appleton-Century-Croft, 1967), 16–22.
14. Ibid., 20.
15. See the description of this process in Klob and Whishaw, *Fundamentals of Human Neuropsychology,* 334.
16. Saussure, *Course in General Linguistics,* 112.
17. Catherine Belsey, *Critical Practice* (London: Methuen, 1980), 40.
18. Benjamin L. Whorf, *Language, Thought, and Reality,* ed. J. Carroll (Cambridge,

Mass.: MIT Press, 1979), 147. Whorf's hypothesis is quite subtle, and while it is usually read in terms of linguistic relativity, it may be interpreted to imply that individual writers impose their own worldview on the language they use. Such a view George Steiner holds. See *On Difficulty and Other Essays* (New York: Oxford University Press, 1978), chap. 6.

19. P. J. Davis and R. Hersh, *Descartes' Dream: The World according to Mathematics* (London: Penguin, 1988), 231.
20. Ibid., 233.
21. Jakobson, *Six Lectures,* lecture 5.
22. Saussure, *Course in General Linguistics,* 123.
23. Ibid.
24. Ibid., 67.
25. Jakobson, *Six Lectures,* 106.
26. Ibid.
27. Conditions of space, such as distance, were discussed in chapter 4. The prereflected conditions of time are those of units, ratio, and comparison, and they combine into a reflected consciousness of random mass change as a function of relatively invariant mass change.
28. On this point Bohm has an interesting discussion on asynordinate structures. When a structure is asynordinate, it is constituted with different degrees of implication. In such cases "the time order is not in general the primary one that is pertinent." Here Bohm is referring to explicit time. Bohm, *Wholeness,* 152–56.
29. Saussure, *Course in General Linguistics,* 67.
30. Lacan, *Ecrits,* 149.
31. Benveniste, *Problems in General Linguistics,* 47.
32. Saussure, *Course in General Linguistics,* 68.
33. Ibid., 104.
34. Anika Lemaire, *Jacques Lacan,* trans. David Macey (London: Routledge and Kegan Paul, 1977), 23.
35. Saussure, *Course in General Linguistics,* 104.
36. Jakobson, *Six Lectures,* 58.
37. Ibid.
38. Ibid., 61.
39. Ibid.
40. Lev Semenovich Vygotsky, *Thought and Language,* trans. Eugenia Hanfmann and Gertrude Vaker (Cambridge, Mass.: MIT Press, 1981), 111.
41. Ibid., 116.
42. Ibid., 110.

Chapter 12

1. Saussure, *Course in General Linguistics,* 80.
2. Ibid., 123.

3. I have based these comments on Saussure's *Course in General Linguistics* and Jakobson's *Six Lectures.*
4. Roman Jakobson, *On Language,* ed. L. Waugh and M. Monville-Burston (Cambridge, Mass.: Harvard University Press, 1990), 115.
5. Ibid.
6. Ibid.
7. Jakobson and Halle, *Fundamentals of Language,* 74.
8. Ibid.
9. The isomorphic implications for Jakobson's two types of aphasia would suggest that the category terms "similarity" and "contiguity" disorder should be replaced by terms that more accurately describe the dialectical organization of discourse. This is not the place to detour into a close study of aphasic conditions from an isomorphic perspective on discourse; however, one possible avenue for this isomorphic study would seem to be the concepts of context and constituent and how these factors affect the elements of transformation and scale. Jakobson's comments concerning constituent/context relationships do not take into account the semantic level or scale at which a deficiency occurs, or how this scaling factor relates to different areas of the brain that are impaired. If we were to apply the chaos concept of scaling to discourse, we would also have to emphasize the differences of scales along with recursive similarities of the different scaling levels of discourse. For example, is a deficiency in the "capacity for naming" an unscaled, general deficiency (implied by Jakobson) brought about by a general reliance on context to the general detriment of any constituent? Or is the problem one at the level of the sentence, where the context of the sentence takes precedence over the subject of a sentence? Surely the factor of scale is important here. And how is this problem different from dyslexia, in which constituent letters in the context of a word can also be seen as a deficiency in the "capacity for naming"? Clearly these two cases of aphasia are different in terms of scale (and correspondingly in terms of neurology), but they share a basic underlying difficulty in the capacity for naming. Jakobson's general model, however, does not allow for this kind of recursive scaling. Such questions as these arise from an isomorphic perspective of discourse and will need much more thorough investigation before they can be answered.
10. Jakobson and Halle, *Fundamentals of Language,* 86.
11. See Gleick, *Chaos,* 101.
12. On this point see Saussure, *Course in General Linguistics,* chap. 1, part 1.
13. To anticipate a Platonic response that nouns are only images or copies of things: it does not matter if the physical referent is one perceived at the moment of designation or a remembered object. Within the framework of the system in general of meaning, both are system types that are then identified with the system type of a linguistic system.
14. For instance see Lacan, *Ecrits,* chap. 5.
15. Claude Lévi-Strauss, *Structural Anthropology,* trans. C. Jacobson and B. Grundfest Schoepf (Middlesex: Penguin, 1968), 210.
16. Ibid.

17. For a clearly described history of rhetoric see Renato Barilli, *Rhetoric,* trans. G. Menozzi (Minneapolis: University of Minnesota Press, 1989).
18. Gunther Kress and Robert Hodge, *Language as Ideology* (London: Routledge and Kegan Paul, 1979), chap. 2, "Transformations and Truth."
19. Ibid., 28.
20. Michel Foucault, *This Is Not a Pipe,* trans. James Harkness (Berkeley and Los Angeles: University of California Press, 1982).

Bibliography

Aitchison, Jean. *The Articulate Mammal.* London: Hutchinson, 1976.

Allan, Keith. *Linguistic Meaning.* Vol. 1. London: Routledge and Kegan Paul, 1986.

Anokhin, Peter K. *Biology and Neurophysiology of the Conditioned Reflex and Its Role in Adaptive Behavior.* Trans. S. A. Corson et al. Oxford: Pergamon, 1974.

Ayer, Alfred J. *Wittgenstein.* Chicago: University of Chicago Press, 1986.

Barilli, Renato. *Rhetoric.* Trans. G. Menozzi. Minneapolis: University of Minnesota Press, 1989.

Bateson, Gregory. *Mind and Nature: A Necessary Unity.* Toronto: Bantam, 1979.

Belsey, Catherine. *Critical Practice.* London: Methuen, 1980.

Benveniste, Emile. *Problems in General Linguistics.* Trans. M. E. Meek. Coral Gables, Fla.: University of Miami Press, 1971.

Birch, Charles. *On Purpose.* Sydney: New South Wales University Press, 1990.

Bohm, David. *Unfolding Meaning.* London: Ark, 1985.

———. *Wholeness and the Implicate Order.* London: Ark, 1983.

Bohm, David, and Basil J. Hiley. *The Undivided Universe.* London: Routledge, 1995.

Bohm, David, and F. David Peat. *Science, Order, and Creativity.* London: Routledge, 1989.

Bray, J., et al., eds. *Lecture Notes on Human Physiology.* Oxford: Blackwell Scientific Publications, 1986.

Bronowski, Jacob. *The Origins of Knowledge and Imagination.* New Haven, Conn.: Yale University Press, 1978.

Brown, H. *Rationality.* London: Routledge, 1988.

Calvin, W. H. *The Throwing Madonna: Essays on the Brain.* New York: Bantam, 1991.

Capra, Fritjof. *The Tao of Physics.* London: Flamingo, 1992.

Changeux, J. *Neuronal Man.* Trans. L. Garey. New York: Oxford University Press, 1986.

Chomsky, Noam. *Language and Mind.* New York: Harcourt Brace Jovanovich, 1972.

———. *Language and Problems of Knowledge.* Cambridge, Mass.: MIT Press, 1985.

———. *Language and Responsibility.* New York: Pantheon, 1979.

———. *Syntactic Structures.* The Hague: Mouton, 1985.

Churchland, Patricia S. *Neurophilosophy.* Cambridge, Mass.: MIT Press, 1986.

Cobb, John. "God in the Age of Science." An address to the Anglican Church Hunters Hill, Sydney, 3 September 1992.

The Concise Yoga Vasistha. Trans. Swami Venkatesananda. Albany: State University of New York Press, 1984.

Crick, Francis. *The Astonishing Hypothesis: The Scientific Search for the Soul.* London: Simon and Schuster, 1994.

Crick, Francis, and Christof Koch. "The Problem of Consciousness." *Scientific American,* September 1992, 111–17.

Damasio, Antonio R., and Hanna Damasio. "Brain and Language." *Scientific American,* September 1992, 63–71.

Davis, L. *Resisting Novels.* New York: Methuen, 1987.

Davis, P. J., and R. Hersh. *Descartes' Dream: The World According to Mathematics.* London: Penguin, 1988.

deGroot, J., and J. G. Chusid. *Correlative Neuroanatomy.* New York: Prentice-Hall, 1985.

Dennett, Daniel C. *Consciousness Explained.* London: Allen Lane, Penguin, 1991.

Denton, Derek. *The Pinnacle of Life: Consciousness and Self-awareness in Humans and Animals.* Sydney: Allen and Unwin, 1993.

Derrida, Jacques. *Margins of Philosophy.* Trans. Alan Bass. Chicago: University of Chicago Press, 1982.

———. *Of Grammatology.* Trans. Gayatri Chakravorty Spivak. Baltimore: John Hopkins University Press, 1980.

———. *Positions.* Trans. Alan Bass. Chicago: University of Chicago Press, 1981.

———. *Speech and Phenomena.* Trans. David B. Allison. Evanston, Ill.: Northwestern University Press, 1989.

———. *Writing and Difference.* Trans. Alan Bass. Chicago: University of Chicago Press, 1978.

Docker, J. *In a Critical Condition.* Melbourne: Penguin, 1984.

Eco, Umberto. *The Role of the Reader.* London: Hutchinson, 1981.

Eco, Umberto, and S. Hall. "Crisis? What Crisis?" *Voices: Writers and Politics.* Channel 4 Television Series. Ed. B. Bourne. Nottingham: Spokesman, 1987.

Eco, Umberto, M. Santambrogio, and P. Violi, eds. *Meaning and Mental Representations.* Bloomington: Indiana University Press, 1988.

Ellis, John. *Against Deconstruction.* Princeton, N.J.: Princeton University Press, 1989.

Fish, Stanley. *Doing What Comes Naturally.* Durham, N.C.: Duke University Press, 1989.

Foucault, Michel. *This Is Not a Pipe.* Trans. James Harkness. Berkeley and Los Angeles: University of California Press, 1982.

Fowler, Roger. *Linguistics and the Novel.* London: Metheun, 1979.

Freadman, A. "On Being Here and Still Doing It." In *The Foreign Bodies Papers.* Sydney: Local Consumption, 1981.

Freud, Sigmund. *The Interpretations of Dreams.* Ed. and trans. James Strachey. Middlesex: Penguin, 1986.

———. "The Unconscious." In *Metapsychology: The Theory of Psychoanalysis.* Trans. J. Strachey. Middlesex: Penguin, 1984.

Frisby, John P. *Seeing Illusion, Brain, and Mind.* New York: Oxford University Press, 1979.

Froula, C. "Quantum Physics/Post-modern Metaphysics: The Nature of Jacques Derrida." *Western Humanities Review* 39 (1985): 286–313.

Gleick, James. *Chaos: Making a New Science.* London: Cardinal, 1989.

Globus, Gordon D. "Three Holonomic Approaches to the Brain." In *Quantum Implications: Essays in Honour of David Bohm,* ed. Basil J. Hiley and F. David Peat. London: Routledge, 1988.

Greenberg, M. *Corneille, Classicism, and the Ruses of Symmetry*. London: Cambridge University Press, 1986.

Greimas, A. J., and J. Courtes. *Semiotics and Language: An Analytical Dictionary.* Trans. Larry Crist et al. Bloomington: Indiana University Press, 1982.

Grof, Stanislav. *The Holotropic Mind.* San Francisco: Harper, 1990.

Gruber, B., and R. S. Millman. *Symmetries in Science.* New York: Plenum Press, 1979.

Halliday, Michael. *Halliday: System and Function in Language.* (Oxford: Oxford University Press, 1981.)

———. "A Short Introduction to Functional Grammar." Sydney University Linguistics Department. Unpublished manuscript, 1983.

Hanfling, O. *Wittgenstein's Later Philosophy.* London: Macmillan, 1989.

Hayles, N. Katherine. "Chaos as Orderly Disorder: Shifting Grounds in Contemporary Literature and Science." *New Literary History* 20 (1990): 305–22.

———, ed. *Chaos and Order.* Chicago: University of Chicago Press, 1991.

———. *Chaos Bound: Orderly Disorder in Contemporary Literature and Science.* London: Cornell University Press, 1990.

Heidegger, Martin. *Being and Time.* Trans. J. Macquarrie and E. Robinson. Southhampton: Basil Blackwell, 1983.

Higgins, M., and C. M. Raphael, eds. *Reich Speaks of Freud.* New York: Noonday, 1967.

Hiley, Basil J., and F. David Peat, eds. *Quantum Implications: Essays in Honour of David Bohm.* London: Routledge, 1988.

Hill, T. E. *The Concept of Meaning.* London: George Allen and Unwin, 1974.

Hogan, John. "Last Words of a Quantum Heretic." *New Scientist,* 27 February 1993, 38–42.

Hollis, M. "The Limits of Rationality." In *Rationality,* ed. B. Wilson. Oxford: Basil Blackwell, 1981.

Jackendoff, R. *Semantics and Cognition.* Cambridge, Mass.: MIT Press, 1990.

Jakobson, Roman. *On Language.* Ed. L. Waugh and M. Monville-Burston. Cambridge, Mass.: Harvard University Press, 1990.

———. *Six Lectures on Sound and Meaning.* Cambridge, Mass.: MIT Press, 1978.

Jakobson, Roman, and Morris Halle. *Fundamentals of Language.* The Hague: Mouton, 1980.

Jakobson, Roman, and K. Pomorska. *Dialogues.* Trans. C. Hubert. Cambridge, Mass.: MIT Press, 1983.

Johnson, Mark. *The Body in the Mind.* Chicago: University of Chicago Press, 1987.

Johnson-Laird, P. N. *The Computer and the Mind.* London: Fontana, 1988.

Klob, B., and I. Q. Whishaw. *Fundamentals of Human Neuropsychology.* 3d ed. New York: W. H. Freeman, 1990.

Korzybski, Alfred. *Science and Sanity.* Lakeville, Conn.: Institute of General Semantics, 1948.

Kress, Gunther, and Robert Hodge. *Language as Ideology*. London: Routledge and Kegan Paul, 1979.

Kristeva, Julia. *Desire in Language*. Ed. Leon Roudiez, trans. Thomas Gora, Alice Jardine, and Leon Roudiez. New York: Columbia University Press, 1980.

Kuhn, Thomas S. *The Structure of Scientific Revolutions*. 2d ed. Chicago: University of Chicago Press, 1970.

Lacan, Jacques. *Ecrits: A Selection*. Trans. Alan Sheridan. London: Tavistock, 1985.

Langer, Monica M. *Merleau-Ponty's Phenomenology of Perception*. London: Macmillan, 1989.

Laplanche, J., and Pontalis, J.-B. *The Language of Psycho-analysis*. Trans. D. Nicholson-Smith. New York: Norton, 1973.

Lakoff, George. "Cognitive Semantics." In *Meaning and Mental Representation,* ed. Umberto Eco, M. Santambrogio, and P. Violi. Bloomington: Indiana University Press, 1988.

Lawson, H. *Reflexivity*. La Salle, Ill.: Open Court, 1985.

Lemaire, Anika. *Jacques Lacan*. Trans. David Macey. London: Routledge and Kegan Paul, 1977.

Lentricchia, Frank, and Thomas McLaughlin, eds. *Critical Terms for Literary Study*. Chicago: University of Chicago Press, 1990.

Lévi-Strauss, Claude. *Structural Anthropology*. Trans. C. Jacobson and B. Grundfest Schoepf. Middlesex: Penguin, 1968.

Lewin, Roger. *Complexity: Life at the Edge of Chaos*. London: J. M. Dent, 1993.

Lloyd, G. *The Man of Reason*. London: Methuen, 1986.

Lukes, S. "Some Problems of Rationality." In *Rationality,* ed. B. Wilson. Oxford: Basil Blackwell, 1981.

MacCabe, C., ed. *The Talking Cure*. London: Macmillan, 1986.

Margenau, Henry. *The Nature of Physical Reality*. Woodbridge: Conn. Ox Bow Press, 1977.

Martin, Graham Dunstan. *Shadows in the Cave*. London: Arkana, 1990.

Matte-Blanco, Ignacio. *Thinking, Feeling, and Being*. Ed. D. Tuckett. London: Routledge, 1988.

———. *The Unconscious as Infinite Sets*. London: Duckworth, 1975.

Maturana, Humberto R. "Autopoiesis: Reproduction, Heredity and Evolution." In *Autopoiesis, Dissipative Structures, and Spontaneous Social Orders,* ed. Milan Zeleny. Boulder, Colo.: Westview Press, 1980.

———. "Biology of Language." In *Psychology and Biology of Language and Thought,* ed. George A. Millar and E. Lenneberg. New York: Academic Press, 1978.

Maturana, Humberto R., and Varela, Francisco J. *Autopoiesis and Cognition: The Realisation of the Living*. Dordrecht, Holland: Reidel, 1980.

———. *The Tree of Knowledge*. Trans. R. Paolucci. Boston: Shambhala, 1992.

McGinn, Colin. *Wittgenstein on Meaning*. Oxford: Basil Blackwell, 1987.

McLeish, John. *Soviet Psychology*. London: Methuen, 1975.

Merleau-Ponty, Maurice. *Consciousness and the Acquisition of Language*. Evanston, Ill.: Northwestern University Press, 1973.

Millar, George A. *Psychology: The Science of Mental Life*. London: Hutchinson, 1966.

Millar, George A., and E. Lenneberg, eds. *Psychology and Biology of Language and Thought*. New York: Academic Press, 1978.

Miller, Alice. *For Your Own Good*. Trans. H. and H. Hannum. New York: Noonday Press, 1991.

Morris, Charles. *Writings on the General Theory of Signs*. The Hague: Mouton, 1971.

Neisser, Ulrich. *Cognitive Psychology*. New York: Appleton-Century-Croft, 1967.

Newman, C. *The Post-Modern Aura*. Evanston, Ill.: Northwestern University Press, 1985.

Nietzsche, Friedrich. *Beyond Good and Evil*. Trans. R. J. Hollingdale. Middlesex: Penguin, 1987.

Ogden, C. K., and I. A. Richards. *The Meaning of Meaning*. London: Ark, 1985.

Pagels, Heinz R. *The Cosmic Code*. London: Penguin, 1994.

———. *Perfect Symmetry*. New York: Bantam, 1991.

Pavlov, Ivan V. *Lectures on Conditioned Reflexes*. Vol. 1. Trans. W. Horsley Gantt. London: Lawrence and Wishart, 1963.

Peat, F. David. *Synchronicity: The Bridge between Mind and Matter*. Toronto: Bantam, 1988.

Penrose, Roger. *The Emperor's New Mind*. London: Vintage, 1990.

Piaget, Jean. "Language and Thought from a Genetic Point of View." In *Language in Thinking*, ed. P. Adams. Middlesex: Penguin, 1973.

Pike, K. L. *Language in Relation to a Unified Theory of the Structure of Human Behavior*. The Hague: Mouton, 1967.

Place, U. T. "The Brain and Consciousness." In *Introductory Readings of Metaphysics*, ed. R. Taylor. Englewood Cliffs, N.J.: Prentice-Hall, 1978.

Pollock, T. C. *A Theory of Meaning Analysed: A Critique of I. A. Richards' Theory of Language and Literature*. General Semantics Monographs, no. 111. Lakeville, Conn.: Institute of General Semantics, 1961.

Popper, Karl, and John C. Eccles. *The Self and Its Brain*. London: Routledge, 1986.

Pribram, Karl H. *Languages of the Brain: Experimental Paradoxes and Principles in Neuropsychology*. New York: Brandon House, 1977.

Prigogine, I., and I. Stengers. *Order Out of Chaos*. London: Flamingo, 1990.

Putnam, Hilary. *Renewing Philosophy*. Cambridge, Mass.: Harvard University Press, 1992.

Reich, Wilhelm. *The Mass Psychology of Fascism*. Trans. V. R. Carfagno. New York: Farrar, Straus and Giroux, 1970.

Riodan, M., and Schramm, D. *The Shadows of Creation: Dark Matter and the Structure of the Universe*. Oxford: Oxford University Press, 1993.

Rosen, Joe. *Symmetry Discovered*. London: Cambridge University Press, 1975.

Sass, Louis A. *Madness and Modernism*. New York: Basic Books, 1992.

Saussure, Ferdinand de. *Course in General Linguistics*. Ed. Charles Bally and Albert Sechehaye, trans. Wade Baskin. Glasgow: Fontana/Collins, 1978.

Schaff, A. *Language and Cognition*. Trans. Olgierd Wojtasiewicz. New York: McGraw-Hill, 1973.

Searle, John R. *Minds, Brains, and Science*. Cambridge, Mass.: Harvard University Press, 1984.

———. *The Rediscovery of the Mind*. Cambridge, Mass.: MIT Press, 1992.

———. *Speech Acts*. Cambridge: Cambridge University Press, 1985.

Shalom, Albert. *The Mind/Body Conceptual Framework and the Problem of Personal Identity*. Atlantic Highlands, N.J.: Humanities Press International, 1989.

Sheldrake, Rupert. *The Presence of the Past.* London: Fontana, 1988.

———. *The Rebirth of Nature.* London: Rider, 1990.

Sokolov, Y. N. *Perception and the Conditioned Reflex.* Trans. S. W. Waydenfeld, London: Pergamond, 1963.

Steiner, George. *On Difficulty and Other Essays.* New York: Oxford University Press, 1978.

Stewart, I. *Does God Play Dice? The Mathematics of Chaos.* Oxford: Basil Blackwell, 1989.

Storr, A., ed. *Jung.* London: Fontana, 1983.

Tallis, R. *Not Saussure: A Critique of Post-Saussurian Literary Theory.* London: Macmillan, 1988.

The Upanishads. Trans. Swami Prabhavananda and Fredrick Manchester. Hollywood, Calif.: Vedanta, 1983.

Varela, Francisco J. *Principles of Biological Anatomy.* Vol. 2. New York: North Holland, 1979.

Varela, Francisco J., J. Evan Thompson, and Eleanor Rosch. *The Embodied Mind.* Cambridge, Mass.: MIT Press, 1992.

Volosinov, V. N. *Marxism and the Philosophy of Language.* Trans. L. Matejka and I. R. Titunik. New York: Seminar Press, 1973.

Vygotsky, Lev Semenovich. *Mind in Society.* Ed. M. Cole et al. Cambridge, Mass.: Harvard University Press, 1978.

———. *Thought and Language.* Trans. Eugenia Hanfmann and Gertrude Vakar. Cambridge, Mass.: MIT Press, 1981.

Wells, Harry K. *Pavlov and Freud: I Ivan P. Pavlov Towards a Scientific Psychology and Psychiatry.* London: Lawrence and Wishart, 1956.

Whorf, Benjamin L. *Language, Thought, and Reality.* Ed. J. Carroll. Cambridge, Mass.: MIT Press, 1979.

Wilber, Ken, ed. *Quantum Questions.* Boston: Shambhala, 1985.

Winograd, Terry, and Fernando Flores. *Understanding Computers and Cognition.* Reading, Mass.: Addison-Wesley, 1988.

Wittgenstein, Ludwig. *Philosophical Investigations.* Trans. G. E. M. Anscombe. Oxford: Basil Blackwell, 1989.

———. *Tractatus Logico-Philosophicus.* Trans. D. F. Pears and B. F. McGuinness. London: Routledge and Kegan Paul, 1961.

Wood, D., and R. Bernasconi, eds. *Derrida and Differance.* Evanston, Ill.: Northwestern University Press, 1968.

Zeleny, Milan. "Autopoiesis: A Paradigm Lost?" In *Autopoiesis, Dissipative Structures, and Spontaneous Social Orders,* ed. M. Zeleny, AAAS Selected Symposia Series. Boulder, Colo.: Westview Press, 1980.

Index